Kotlin 开发进阶

[美] 米洛什·瓦西奇　著

张　博　译

清华大学出版社

北　京

内容简介

本书详细阐述了与 Kotlin 相关的基本解决方案，主要包括开启 Android 之旅、构建和运行应用程序、屏幕、连接屏幕流、观感、权限、与数据库协同工作、Android 偏好设置、Android 中的并发机制、Android 服务、消息机制、后端和 API、性能调优、测试、迁移至 Kotlin、部署应用程序等内容。此外，本书还提供了相应的示例、代码，以帮助读者进一步理解相关方案的实现过程。

本书既可作为高等院校计算机及相关专业的教材和教学参考书，也可作为相关开发人员的自学教材和参考手册。

Copyright © Packt Publishing 2017. First published in the English language under the title *Mastering Android Development with Kotlin.*

Simplified Chinese-language edition © 2019 by Tsinghua University Press. All rights reserved.

本书中文简体字版由 Packt Publishing 授权清华大学出版社独家出版。未经出版者书面许可，不得以任何方式复制或抄袭本书内容。

北京市版权局著作权合同登记号 图字：01-2018-1020

本书封面贴有清华大学出版社防伪标签，无标签者不得销售。
版权所有，侵权必究。侵权举报电话：010-62782989 13701121933

图书在版编目（CIP）数据

Kotlin 开发进阶 /（美）米洛什·瓦西奇著；张博译．—北京：清华大学出版社，2019.10
书名原文：Mastering Android Development with Kotlin
ISBN 978-7-302-53928-5

Ⅰ．①K… Ⅱ．①米… ②张… Ⅲ．①JAVA 语言-程序设计 Ⅳ．①TP312.8

中国版本图书馆 CIP 数据核字（2019）第 224316 号

责任编辑： 贾小红
封面设计： 刘 超
版式设计： 文森时代
责任校对： 马军令
责任印制： 李红英

出版发行： 清华大学出版社
网 址： http://www.tup.com.cn，http://www.wqbook.com
地 址： 北京清华大学学研大厦 A 座 **邮 编：** 100084
社 总 机： 010-62770175 **邮 购：** 010-62786544
投稿与读者服务： 010-62776969，c-service@tup.tsinghua.edu.cn
质量反馈： 010-62772015，zhiliang@tup.tsinghua.edu.cn
印 刷 者： 北京富博印刷有限公司
装 订 者： 北京市密云县京文制本装订厂
经 销： 全国新华书店
开 本： 185mm×230mm **印 张：** 20 **字 数：** 400 千字
版 次： 2019 年 11 月第 1 版 **印 次：** 2019 年 11 月第 1 次印刷
定 价： 109.00 元

产品编号：078394-01

译　者　序

Kotlin 是一种新型语言且具有较好的稳定性，并可在所有 Android 设备上运行，同时还解决了 Java 无法处理的许多问题。Kotlin 为 Android 开发平台引入了许多已被证实的编程概念，使得开发过程变得更加轻松，并可生成更具安全性、表现力和简洁的代码。同时，也希望读者具备开阔的头脑，以及对新技术的渴望之心，这对程序设计学习来说十分有益。

针对于此，本书精心挑选了与 Kotlin 语言相关的进阶开发实例，涉及构建和运行应用程序、屏幕、连接屏幕流、观感、权限、与数据库协同工作、Android 偏好设置、Android 中的并发机制、Android 服务、消息机制、后端和 API、性能调优、测试、迁移至 Kotlin、部署应用程序等内容。这里，我们也建议读者重点考查相关代码，并理解其所执行的任务。除此之外，还需要亲自实现、运行书中的每一个程序。

在本书的翻译过程中，除张博之外，刘璋、刘晓雪、刘祎、张华臻等人也参与了部分翻译工作，在此一并表示感谢。

由于译者水平有限，难免有疏漏和不妥之处，恳请广大读者批评指正。

译　者

前　　言

Android 是最为流行的移动设备平台之一，每年都会有大量的开发人员投入 Android 开发当中，而 Android Framework 允许我们针对移动电话、平板电脑、电视等开发相应的应用程序。之前，全部开发任务仅可通过 Java 完成。近期，谷歌发布了 Kotlin 作为开发人员可用的第二种编程语言。随着 Kotlin 的不断壮大，本书将讨论与 Kotlin 相关的编程知识。

借助于 Kotlin，我们可完成通过 Java 所做的一切事物。本书将向读者展示如何通过 Android 和 Kotlin 创建令人惊奇的应用程序。鉴于 Kotlin 的存在，Android 平台也会有更大的发展空间。在不久的将来，Kotlin 很有可能成为该平台的主要开发语言。

本书内容

第 1 章：开启 Android 之旅。将讨论如何利用 Kotlin 进行 Android 开发，以及如何设置相关工作环境。

第 2 章：构建和运行应用程序。将探讨如何构建和运行项目，其中将会涉及应用程序日志和调试方面的内容。

第 3 章：屏幕。将介绍 UI，并针对应用程序创建第一个屏幕。

第 4 章：连接屏幕流。将描述屏幕流的连接方式，以及如何利用 UI 定义基本的用户交互行为。

第 5 章：观感。将讲解 UI 的主题，并阐述 Android 中与主题相关的基本概念。

第 6 章：权限。将探讨系统权限问题，进而可使用特定的系统功能。

第 7 章：与数据库协同工作。将涉及应用程序的存储机制，包括如何使用 SQLite，随后将生成一个数据库以存储和共享数据。

第 8 章：Android 偏好设置。将解释并非所有数据都将存储至数据库中，一些信息还可存储于共享偏好设置中。本章将对其原因和方式加以讨论。

第 9 章：Android 中的并发机制。将考查 Android 中的并发机制。读者可从中了解到，多项任务可同步执行，Android 也不例外。

第 10 章：Android 服务。将介绍 Android 服务及其应用方式。

第 11 章：消息机制。将讨论在 Android 中，应用程序可监听各种事件，本章将对此给出答案。

第 12 章：后端和 API。将讨论如何连接至远程后端实例进而获取数据。

第 13 章：性能调优。将主要探讨与应用程序的执行速度相关的性能调优问题。

第 14 章：测试。将探讨在应用程序发布之前的测试问题，并考查如何针对应用程序编写测试程序。

第 15 章：迁移至 Kotlin。将主要介绍如何将现有的 Java 代码迁移至 Kotlin 中。

第 16 章：部署应用程序。将讨论应用程序的部署过程，进而实现应用程序的发布任务。

软件和硬件环境

对于本书，读者需要一台能够运行 Microsoft Windows、Linux 或 macOS 的计算机设备，同时还需要安装 Java JDK、Git 版本控制系统和 Android Studio。

当运行本书示例以及读者所编写的代码时，需要一部能够运行 Android 操作系统（版本不低于 5）的手机设备。

适用读者

本书的目标读者是那些想要以一种简单而有效的方式构建良好 Android 应用程序的开发人员。本书假设读者已基本了解 Kotlin，但尚不熟悉 Android 开发。

本书约定

本书通过不同的文本风格区分相应的信息类型。下面通过一些示例对此类风格以及具体含义的解释予以展示。

代码块如下：

```
override fun onCreate(savedInstanceState: Bundle?) {
  super.onCreate(savedInstanceState)
  setContentView(R.layout.activity_main)
```

```
  Log.v(tag, "[ ON CREATE 1 ]")
}
```

命令行输入或输出则采用下列方式表达：

```
sudo apt-get install libc6:i386 libncurse
libstdc++6:i386 lib32z1 libbz2-1.0:i386
```

图标则表示较为重要的说明事项。

图标则表示提示信息和操作技巧。

软件环境和资源下载

读者可访问 http://www.packtpub.com 并通过个人账户下载示例代码文件。另外，在 http://www.packtpub.com/support 中注册成功后，我们将以电子邮件的方式将相关文件发与读者。

读者可根据下列步骤下载代码文件：

（1）利用电子邮件和密码登录或注册我们的网站 www.packtpub.com。

（2）单击 SUPPORT 选项卡。

（3）单击 Code Downloads & Errata。

（4）在 Serach 文本框中输入书名。

（5）选择下载的书籍。

（6）从下拉菜单中选择书籍的购买方式。

（7）单击 Code Download 按钮。

当文件下载完毕后，确保使用下列最新版本软件解压文件夹：

- Windows 系统下的 WinRAR/7-Zip。
- Mac 系统下的 Zipeg/iZip/UnRarX。
- Linux 系统下的 7-Zip/PeaZip。

另外，读者还可访问 GitHub 获取本书的代码包，对应网址为 https://github.com/PacktPublishing/-Mastering-Android-Development-with-Kotlin/branches/all。

此外，读者还可访问 https://github.com/PacktPublishing/以了解丰富的代码和视频资源。

最后，读者还可访问 https://www.packtpub.com/sites/default/files/downloads/Mastering

AndroidDevelopmentwithKotlin_ColorImages.pdf 以下载并查看书中的彩色图像。

读者反馈和客户支持

欢迎读者对本书的建议或意见予以反馈。

对此，读者可向 feedback@packtpub.com 发送邮件，并以书名作为邮件标题。若读者对本书有任何疑问，均可发送邮件至 questions@packtpub.com，我们将竭诚为您服务。

若读者针对某项技术具有专家级的见解，抑或计划撰写书籍或完善某部著作的出版工作，则可访问 www.packtpub.com/authors。

勘误表

尽管我们在最大程度上做到尽善尽美，但错误依然在所难免。如果读者发现谬误之处，无论是文字错误抑或是代码错误，还望不吝赐教。对此，读者可访问 http://www.packtpub.com/ submit-errata，选取对应书籍，单击 Errata Submission Form 超链接，并输入相关问题的详细内容。

版权须知

一直以来，互联网上的版权问题从未间断，Packt 出版社对此类问题异常重视。若读者在互联网上发现本书任意形式的副本，请告知网络地址或网站名称，我们将对此予以处理。关于盗版问题，读者可发送邮件至 copyright@packtpub.com。

问题解答

若读者对本书有任何疑问，均可发送邮件至 questions@packtpub.com，我们将竭诚为您服务。

目　　录

第 1 章　开启 Android 之旅

Kotlin 已经被谷歌正式宣布为 Android 的一级编程语言。本书将探讨为何 Kotlin 对于编程新手来说是一款优秀的开发工具，以及为何高级开发人员首先采用 Kotlin 作为其开发工具。

本章将探讨如何构建工作环境、安装和运行 Android Studio、配置 Android SDK 和 Kotlin。除此之外，本章还将介绍一些较为重要的工具，如 Android Debug Bridge（ADB）。

在开始阶段，读者可初始化一个 Git 存储库以跟踪代码中的变化内容，同时生成一个空项目以使其支持 Kotlin，并添加其他库以供后续操作使用。

在存储库和项目初始化完毕后，我们将考查对应的项目结构，并解释 IDE 生成的每一个文件。最后，我们将创建第一个程序画面，并对其进行考查。

本章主要涉及以下主题：

- 设置 Git 和 Gradle 开发环境。
- 与 Android Manifest 协同工作。
- Android 模拟器。
- Android 工具。

1.1　为何选择 Kotlin

在开始漫长的旅程之前，我们首先需要回答这一问题：为何选择 Kotlin 语言？Kotlin 是由开发 IntelliJ IDEA 的 JetBrains 公司推出的一种新的编程语言。Kotlin 简洁易懂，它像 Java 一样将所有内容编译成字节码，此外也可以编译成 JavaScript 或原生码。

Kotlin 源自业界的专业人士，旨在解决开发人员所面临的各种问题。同时，IntelliJ 还内置了一个 Java-Kotlin 转换工具，并可逐个文件地转换 Java 代码，同时保持一切可顺利工作。

Kotlin 具有互操作性，并支持任何现有的 Java 框架和库。这种互操作性实现了无缝衔接，且不需要封装器和适配器层。

借助于外部支持，Kotlin 还支持诸如 Gradle、Maven、Kobalt、Ant 和 Griffon 等构建系统。

对于我们来说，当前最为重要的任务是理解 Kotlin 如何与 Android 实现完美的协同工作。

Kotlin 涵盖了以下一些较为重要的特性：

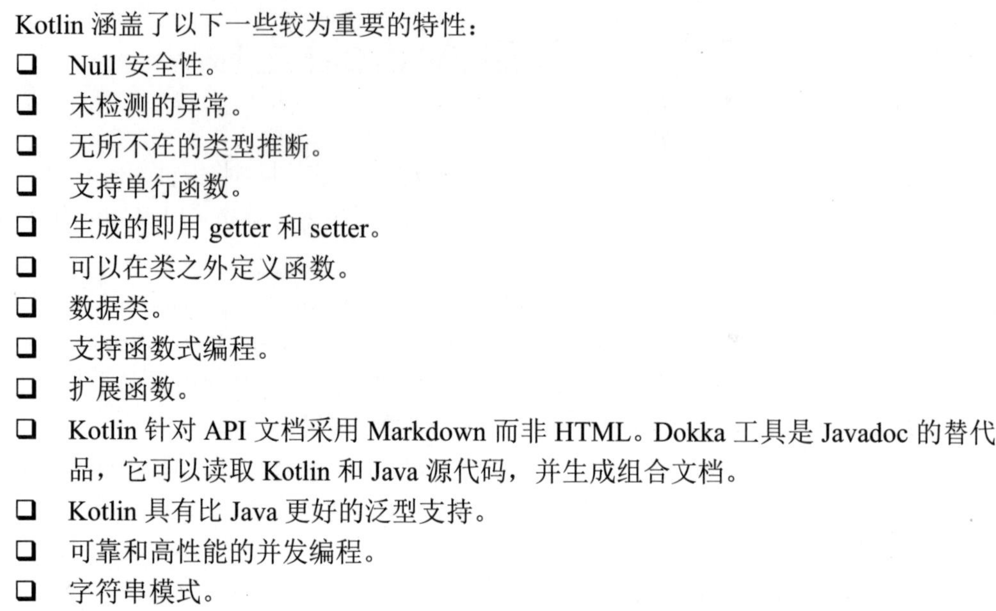

- ❑ Null 安全性。
- ❑ 未检测的异常。
- ❑ 无所不在的类型推断。
- ❑ 支持单行函数。
- ❑ 生成的即用 getter 和 setter。
- ❑ 可以在类之外定义函数。
- ❑ 数据类。
- ❑ 支持函数式编程。
- ❑ 扩展函数。
- ❑ Kotlin 针对 API 文档采用 Markdown 而非 HTML。Dokka 工具是 Javadoc 的替代品，它可以读取 Kotlin 和 Java 源代码，并生成组合文档。
- ❑ Kotlin 具有比 Java 更好的泛型支持。
- ❑ 可靠和高性能的并发编程。
- ❑ 字符串模式。
- ❑ 命名方法参数。

1.2 Android 官方语言——Kotlin

2017 年 5 月 17 日，谷歌宣布将 Kotlin（一种用于 Java 虚拟机的静态类型编程语言）作为编写 Android 应用程序的一级语言。与此同时，Android Studio 的下一个版本（即 3.0 版本，当前版本是 2.3.3）也将对 Kotlin 予以支持，进而在 Kotlin 语言上全面发力。

注意：

Kotlin 仅是一种附加的语言，而不是现有 Java 和 C++语言的替代品（就目前来说）。

1.3 下载和配置 Android Studio

当开发应用程序时，我们需要安装一些相关工具，首先是 IDE。针对于此，我们可采用 Android Studio。Android Studio 针对各种 Android 设备类型提供了快速构建工具。

同时，Android Studio 还提供了专业的代码编辑机制、调试功能以及高性能的开发工具，这是一个较为灵活的构建系统，以使开发人员能够专注于构建高质量的应用程序。

Android Studio 的设置过程需要花费一点时间，在进一步讨论之前，首先需要针对当前操作系统下载 Android Studio，对应网址为 https://developer.android.com/studio/index.html。

下列内容分别针对 macOS、Linux 和 Windows 操作系统列出了具体安装步骤。

1. macOS 操作系统

当在 macOS 操作系统上安装 Android Studio 时，需要执行以下步骤：

（1）打开 Android Studio DMG 文件。

（2）将 Android Studio 拖曳至 Applications 文件夹中。

（3）启动 Android Studio。

（4）选择是否导入之前的 Android Studio 的配置。

（5）单击 OK 按钮。

（6）遵从相关指令，直至 Android Studio 准备就绪。

2. Linux 操作系统

当在 Linux 操作系统上安装 Android Studio 时，需要执行以下步骤：

（1）将下载的存档文件解压缩到应用程序的适当位置。

（2）访问 bin/directory/。

（3）执行/studio.sh。

（4）选择是否导入之前的 Android Studio 的配置。

（5）单击 OK 按钮。

（6）遵从相关指令，直至 Android Studio 准备就绪。

（7）从下拉菜单中选择 Tools | Create Desktop Entry 命令。

注意：

当运行 Ubuntu 的 64 位版本时，需要利用下列命令安装某些 32 位的库：

```
sudo apt-get install libc6:i386 libncurses5:i386
libstdc++6:i386 lib32z1 libbz2-1.0:i386
```

当运行 64 位 Fedora 时，对应命令如下所示。

```
sudo yum install zlib.i686 ncurses-libs.i686 bzip2-libs.i686
```

3. Windows 操作系统

当在 Windows 操作系统上安装 Android Studio 时，需要执行以下步骤：

（1）执行下载后的.exe 文件。

（2）遵从相关指令，直至 Android Studio 准备就绪。

1.4 配置 Android 模拟器

Android SDK 内置了模拟器，进而可运行所开发的应用程序，这也是项目开发过程中不可或缺的工具之一。这里，模拟器的功能旨在模拟某台设备并显示计算机窗口中的全部活动，因而可在没有硬件设备的情况下构建原型、开发和测试。具体来说，模拟器可模拟移动电话、平板电脑、穿戴设备以及电视设备。当然，用户也可设置自己的设备定义，或者采用预定义的模拟器。

模拟器的优点在于速度。在大多数场合下，与实际的硬件设备相比，在模拟器实例上运行应用程序所需的时间更少。

另外，使用模拟器与使用真正的硬件设备一样简单，例如，手势可使用鼠标，输入可采用键盘。

模拟器可实现手机所做的任何事情，用户可轻松地处理来电和短信、指定设备的位置、发送指纹扫描、调整网络速度和状态，设置可以模拟电池的性能。另外，模拟器还配置了虚拟 SD 卡，用户可将实际文件发送至其中。

Android 虚拟设备（Android Virtual Device，AVD）配置用于定义一个模拟器。每个 AVD 实例完全独立于设备；而对于 AVD 的创建和管理，我们可使用 AVD Manager。AVD 配置中涵盖了硬件配置、系统镜像、存储区域、皮肤以及其他重要的属性。

当运行 AVD Manager 时，可执行如下操作：选择 Tools | Android | AVDManager 命令或单击工具栏上的 AVDManager 图标，如图 1.1 所示。

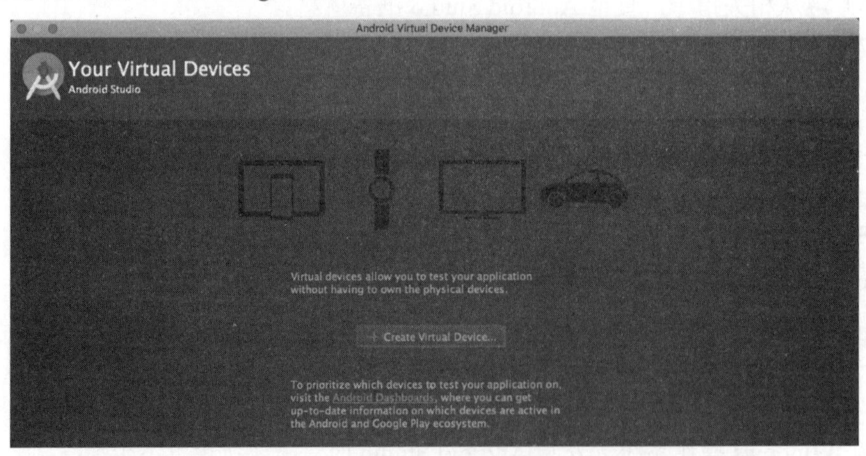

图 1.1

这将显示所定义的全部 AVD。当然，目前尚未定义任何 AVD。

当前，我们可执行下列操作：

- ❑ 创建一个新的 AVD。
- ❑ 编辑现有的 AVD。
- ❑ 删除现有的 AVD。
- ❑ 创建硬件配置。
- ❑ 编辑现有的硬件配置。
- ❑ 删除现有的硬件配置。
- ❑ 导入/导出定义。
- ❑ 启动或终止 AVD。
- ❑ 清空数据并重置 AVD。
- ❑ 访问文件系统中的 AVD .ini 文件和.img 文件。
- ❑ 查看 AVD 配置细节信息。

当获取一个 AVD 实例时，可从头创建一个新的 AVD；或者复制一个现有的 AVD，并根据需要对其进行修改。

1.4.1　创建一个新的 AVD

在 AVD Manager 的 Your Virtual Devices 中，单击 Create Virtual Device 按钮（在 Android Studio 中运行应用程序时，也可执行相同的操作，即单击 Run 图标，并在弹出的 Select Deployment Target 对话框中选择 Create New Emulator），如图 1.2 所示。

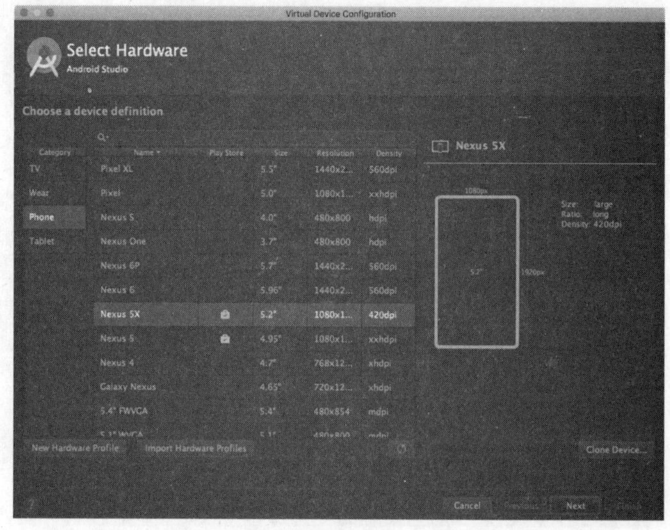

图 1.2

选择硬件配置，随后单击 Next 按钮，如图 1.3 所示。

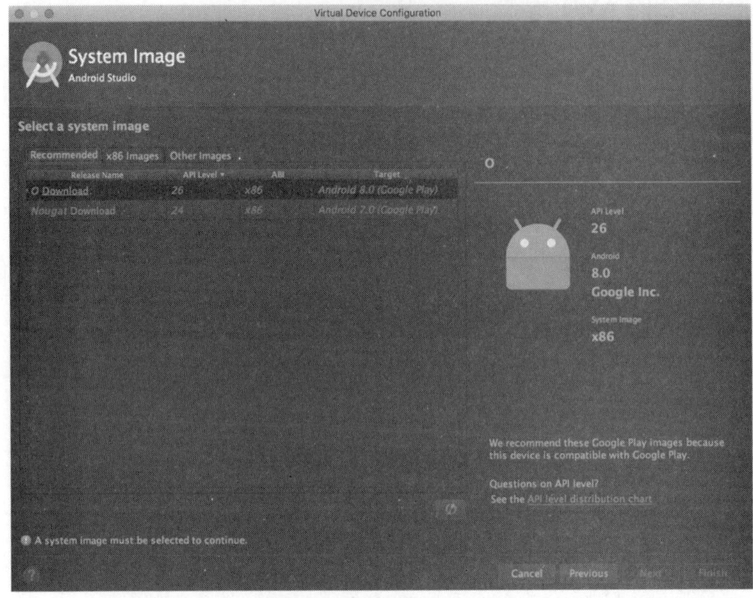

图 1.3

单击 Download 链接即启动下载过程，如图 1.4 所示。

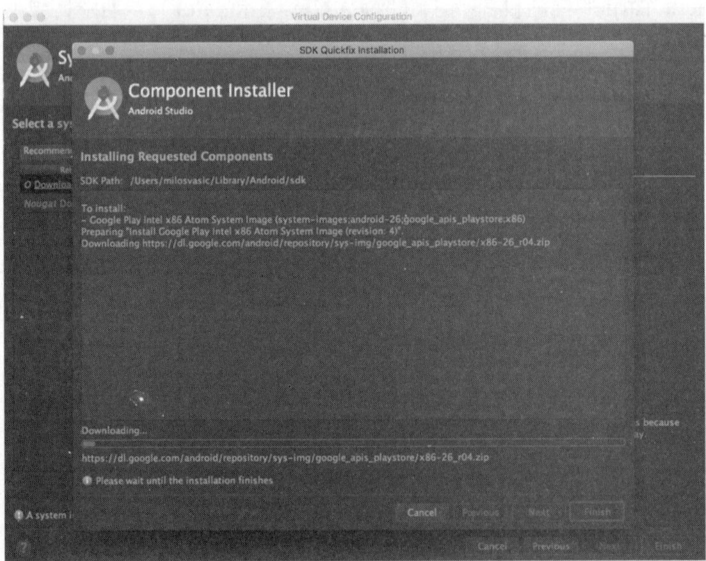

图 1.4

应用程序不能在 API 级别低于应用程序所需的系统镜像上运行。该属性在 Gradle 配置中指定，稍后我们将详细讨论 Gradle。

图 1.5 显示了 Verify Configuration 项。

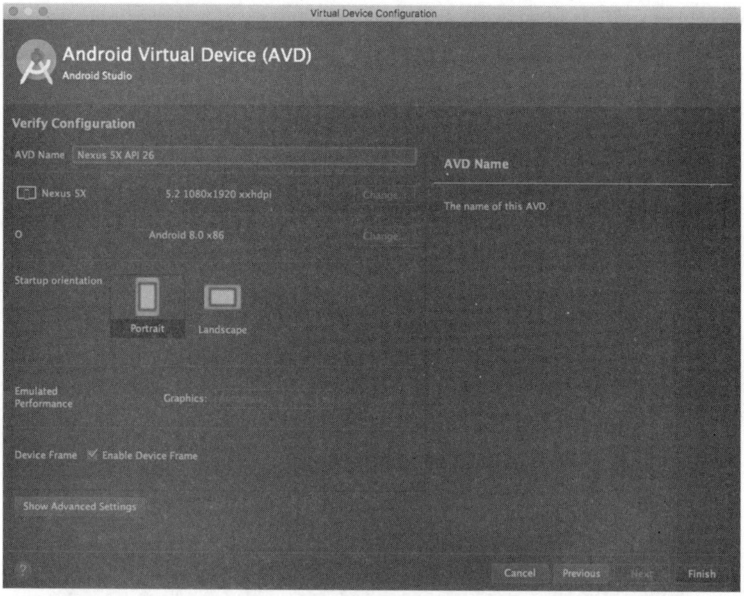

图 1.5

必要时可修改 AVD 属性，并于随后单击 Finish 按钮完成配置向导。此时，新创建的 AVD 将出现于 Select Deployment Target 对话框的 Your Virtual Devices 列表中（取决于配置向导的访问位置），如图 1.6 所示。

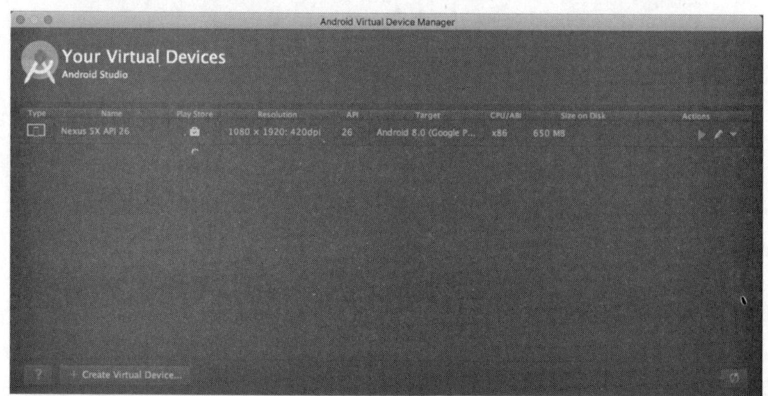

图 1.6

如果需要创建已有的 AVD 副本，则可遵循下列步骤：

（1）打开 AVD Manager，右击 AVD 实例并选择 Duplicate。

（2）在配置向导中修改设置条件，并于随后单击 Finish 按钮。

（3）AVD 的修正版本将出现于 AVD 列表中。

下面将讨论硬件配置的构建过程。当创建一个新的硬件配置时，可在 Select Hardware 中单击 New Hardware Profile 按钮，这将显示 Configure Hardware Profile 窗口，如图 1.7 所示。

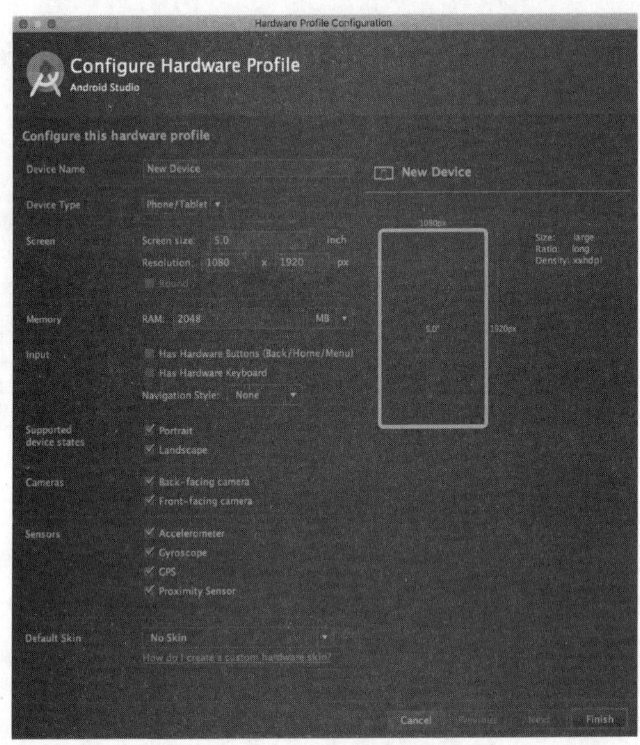

图 1.7

相应地，可根据需要修改 hardware profile 属性并单击 Finish 按钮，随后将显示新创建的 hardware profile。

1.4.2　复制、修改现有的 AVD

当根据已有的硬件配置设置 hardware profile 时，可执行下列步骤：

（1）选择现有的 hardware profile 并单击 Clone Device 按钮。

（2）根据需要更新 hardware profile，单击 Finish 按钮完成配置向导。

（3）配置结果将显示于 hardware profile 列表中。

下面返回 AVD 列表。针对已有的 AVD，可执行下列各项操作：

- 单击 Edit 编辑 AVD。
- 右击并选择 Delete 进而删除 AVD。
- 在 AVD 实例上右击并选择 Show on Disk，进而访问磁盘上的.ini 和.img 文件。
- 当查看 AVD 配置细节信息时，可在 AVD 实例上右击并选择 View Details。

在此基础上，我们返回 hardware profile 列表并执行下列操作：

- 选取 Edit Device 并编辑硬件属性。
- 右击并选取 Delete，进而删除某个硬件配置。

注意：

用户无法编辑或删除预定义的硬件配置。

接下来，可运行、终止模拟器，或者清空其中的数据，具体如下：

- 当运行使用 AVD 的模拟器时，可双击 AVD 或仅选择 Launch。
- 当终止模拟器时，可右击该模拟器并选择 Stop。
- 当清空模拟器的数据并返回其最初的定义状态时，可右击 AVD 并选择 Wipe Data。

接下来将查看与模拟器相关的命令行操作（可使用*-）。

当启用模拟器时，可使用 emulator 命令。下列内容显示了某些基本的命令行语法，进而从终端中启动某个虚拟设备：

```
emulator -avd avd_name [ {-option [value]} ... ]
```

下列内容显示了另一种命令行语法：

```
emulator @avd_name [ {-option [value]} ... ]
```

考查下列示例：

```
$ /Users/vasic/Library/Android/sdk/tools/emulator -avd
Nexus_5X_API_23 -netdelay none -netspeed full
```

这里，在启用模拟器时可指定启动选项，且无法在后续操作中对其进行再次设置。

当查看有效的 AVD 列表时，可使用下列命令：

```
emulator -list-avds
```

对应结果表示为源自 Android 主目录的 AVD 名称列表。通过设置 ANDROID_SDK_

HOME 环境变量，还可重载默认的主目录。

不难发现，终止模拟器就像关闭一个窗口那样简单。

**注意：**

还可从 Android Studio UI 中运行 AVD。

1.5　Android 调试桥

当访问设备时，可在终端中执行 Android 调试桥（adb）命令，本节将讨论一些较为常见的命令。

下列命令将列出所有的设备：

```
adb devices
```

控制台输出结果如下：

```
List of devices attached
emulator-5554 attached
emulator-5555 attached
```

下列命令将通过 shell 访问设备：

```
adb shell
```

下列命令将访问特定的设备实例：

```
adb -s emulator-5554 shell
```

其中，-s 表示为设备源。

下列命令将在设备间复制一个文件：

```
adb pull /sdcard/images ~/images
adb push ~/images /sdcard/images
```

下列命令将卸载一个应用程序：

```
adb uninstall <package.name>
```

另外，adb 中最重要的一个特性是可通过远程登录进行访问。例如，可通过 telnet localhost 5554 连接模拟器设备，并使用 quit 或 exit 命令结束会话。

下列内容展示了一些 adb 应用。

- 连接设备：

```
telnet localhost 5554
```

- ❑ 调整电源级别：

```
power status full
power status charging
```

- ❑ 模拟通话：

```
gsm call 223344556677
```

- ❑ 发送 SMS：

```
sms send 223344556677 Android rocks
```

- ❑ 设置地理位置：

```
geo fix 22 22
```

注意：

利用 adb，还可获取屏幕截屏或者录制视频。

1.6　其他重要工具

本节将介绍 Android 开发过程中所需的一些其他工具，具体如下：

- ❑ adb dumpsys：当获取与系统相关的信息并运行某个应用程序时，可使用 adb dumpsys 命令；当获取内存状态时，可使用 adb shell dumpsys meminfo <package.name>命令。
- ❑ adb shell procrank：该命令将列出全部应用程序以查看其内存使用状态。注意，该命令仅连接于模拟器，且无法在实际设备上工作。针对这一功能，还可使用 adb shell dumpsys meminfo 命令。
- ❑ 对于电池消耗量，可使用 adb shell dumpsys batterystats--charged <package-name> 命令。
- ❑ Systrace 命令通过获取并显示运行时间，进而分析当前应用程序的性能。

当应用程序出现故障时，即可看出 Systrace 命令的重要性。

当使用该命令时，需要安装、配置 Python，下面将尝试使用 Systrace 命令。

当从 UI 中对其进行访问时，可打开 Android Studio 中的 Android Device Monitor，并选择 Monitor，如图 1.8 所示。

图 1.8

某些时候，也可从终端（命令行）中更加方便地对其加以访问。

注意：
取决于运行于设备上的 Android 版本，Systrace 工具包含了不同的命令行选项。

接下来考查一些相关示例。

其常规应用主要包括：

```
$ python systrace.py [options] [category1] [category2] ... [categoryN]
```

- 在 Android 4.3 及其以上版本：

```
$ python systrace.py --time=15 -o my_trace_001.html
sched gfx view wm
```

- 在 Android 4.2 及其之前的版本：

```
$ python systrace.py --set-tags gfx,view,wm
$ adb shell stop
$ adb shell start
$ python systrace.py --disk --time=15 -o my_trace_001.html
```

最后一个较为重要的命令是 sdkmanager。该命令可查看、安装、更新、卸载 Android SDK 数据包，且位于 android_sdk/tools/bin/。

下面考查一些常见的应用示例。

下列命令将列出所安装的数据包：

```
sdkmanager --list [options]
```

下列命令将安装数据包：

```
sdkmanager packages [options]
```

用户可发送源自--list 命令的数据包。

下列命令用于卸载数据包：

```
sdkmanager --uninstall packages [options]
```

下列命令用于更新数据包：

```
sdkmanager --update [options]
```

此外，Android 中还包含了其他一些可用工具，上述内容仅列出了一些较为重要的工具。

1.7　初始化 Git 存储库

前述内容安装了 Android Studio 并讨论了一些较为重要的 SDK 工具；此外，我们还学习了如何处理运行代码的模拟器设备。本节将开始着手建立相关项目，并开发一个备忘录小应用程序，这也是较为常见的工具。该项目命名为 Journaler，可生成相关提示内容和待办事项并同步至后端。

开发过程中的第一步是初始化 Git 存储库。这里，Git 将作为代码的版本系统，用户可以决定是否将 GitHub、BitBucket 等用作远程 Git 实例。接下来将创建远程存储库并使其 URL 和凭证处于就绪状态。

下列命令将访问包含当前项目的目录：

```
Execute: git init
```

对应的控制台输出结果如下：

```
Initialized empty Git repository in <directory_you_choose/.git>
```

至此，存储库初始化完毕。

下面将添加第一个文件——vi notes.txt。

向 notes.txt 文件中输入一些内容并保存该文件。

随后，执行 git add 并添加所有的相关文件。

接下来运行下列命令：

```
git commit -m "Journaler: First commit"
```

对应的控制台输出结果如下：

```
[master (root-commit) 5e98ea4] Journaler: First commit
1 file changed, 1 insertion(+)
create mode 100644 notes.txt
```

如前所述，包含凭证的远程 Git 存储库 url 已处于就绪状态，将 url 复制至粘贴板中，并执行下列命令：

```
git remote add origin <repository_url>
```

这设置了新的远程实例，接下来执行下列命令：

```
git remote -v
```

这将验证新的远程 URL，随后执行下列命令：

```
git push -u origin master
```

其间，如果对凭证予以询问，可输入当前凭证并按 Enter 键进行确认。

1.8　创建 Android 项目

在初始化了代码存储库后，本节将创建一个项目。对此，启动 Android Studio 并执行下列操作步骤：

（1）启动新的 Android Studio Project 或选择 File | New | New Project 命令。

（2）创建 New Project，将显示如图 1.9 所示的窗口，随后填充相关的应用程序信息。

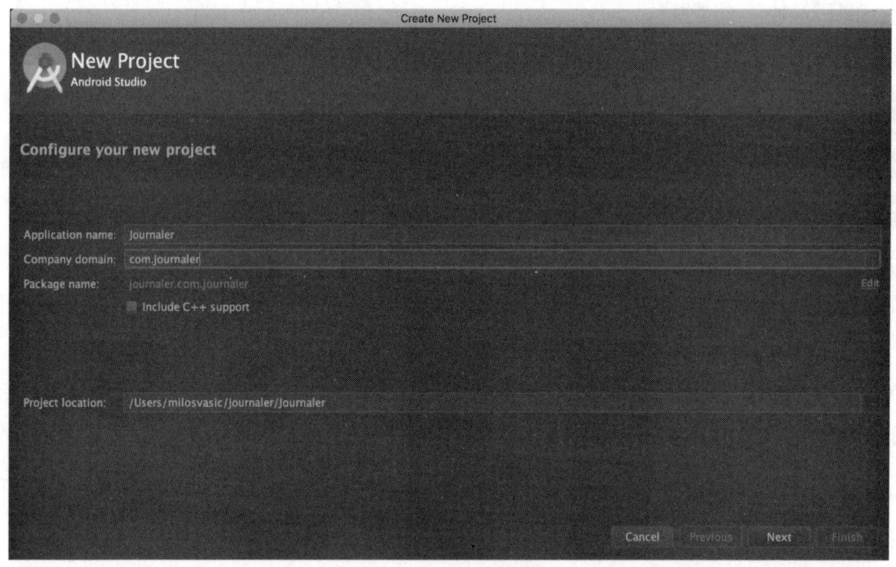

图 1.9

（3）单击 Next 按钮。

（4）选中 Phone and Tablet 复选框，并选取 Android 5.0 作为最低 Android 版本，如图 1.10 所示。

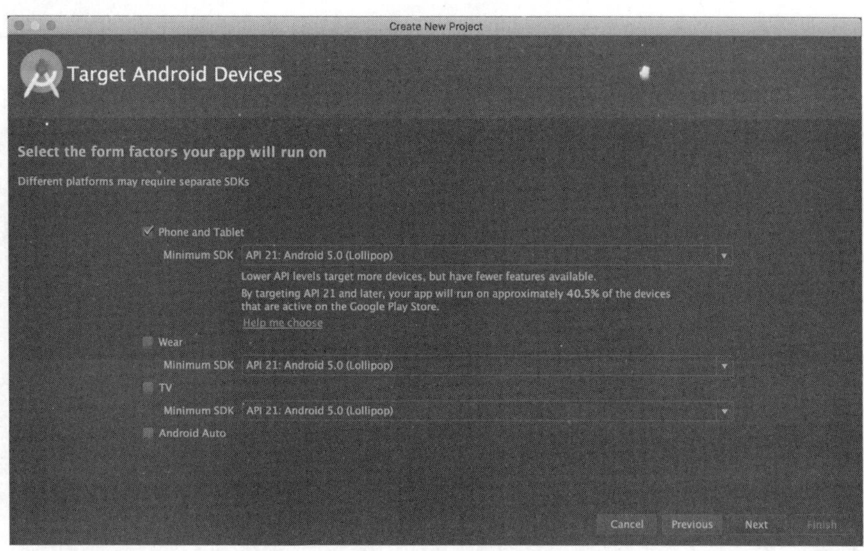

图 1.10

（5）单击 Next 按钮。

（6）选择 Add No Activity 并单击 Finish 按钮，如图 1.11 所示。

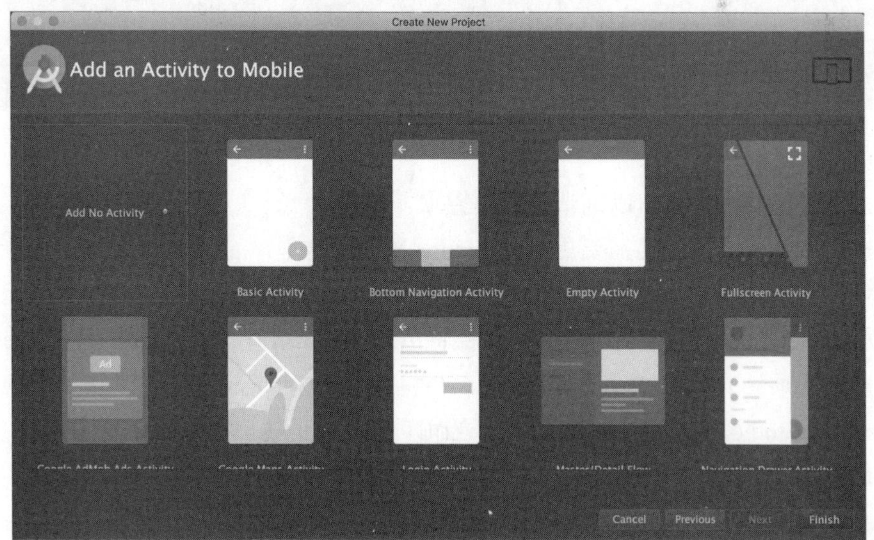

图 1.11

稍作等待，直至项目创建完毕。

其间，用户可能会看到一条与 Unregistered VCS root detected 相关的消息。对此，单击 add root 或选择 Preferences | Version Control，随后从列表中选取我们的 Git 存储库并单击“+”图标，如图 1.12 所示。

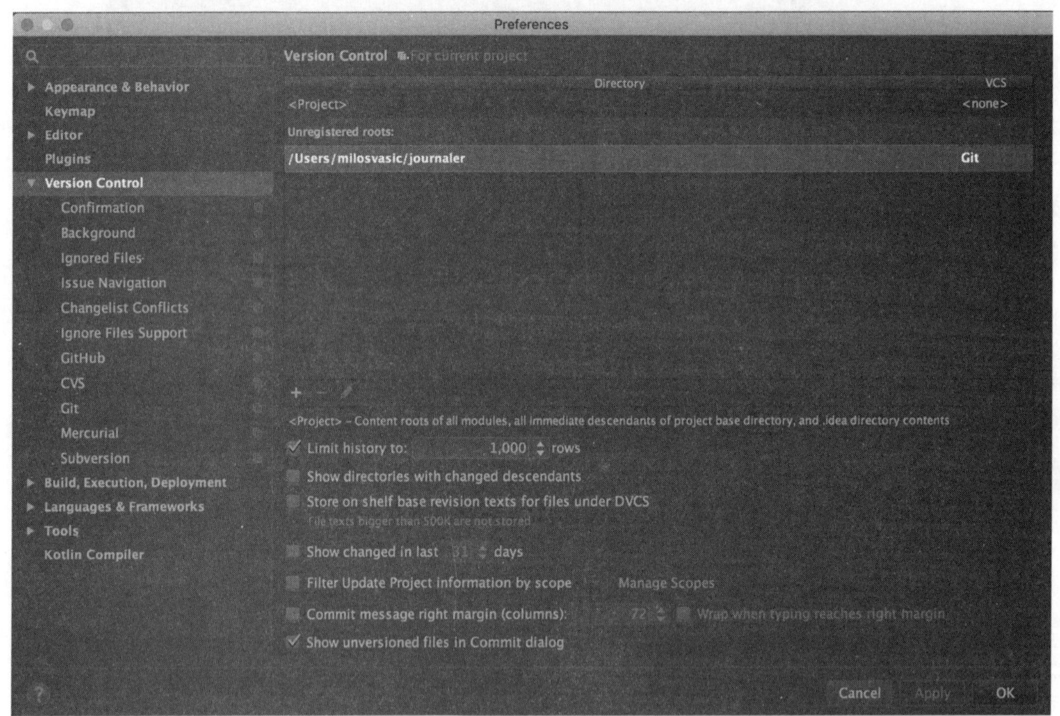

图 1.12

单击 Apply 和 OK 按钮以确认全部设置项。

在提交和推送之前，还需要更新.gitignore 文件。这里，.gitignore 文件的功能是忽略默写文件，例如编辑器备份文件、构建产品或不希望提交至存储库的一些本地配置覆写内容。如果不匹配.gitignore 规则，这些文件将出现在 Git 状态输出的 untracked files 部分。

打开位于项目 root 目录下的.gitignore 文件并对其进行编辑。当对其进行访问时，可单击 Android Studio 左侧的 Project 并展开当前项目，并于随后在下拉菜单中选择 Project，如图 1.13 所示。

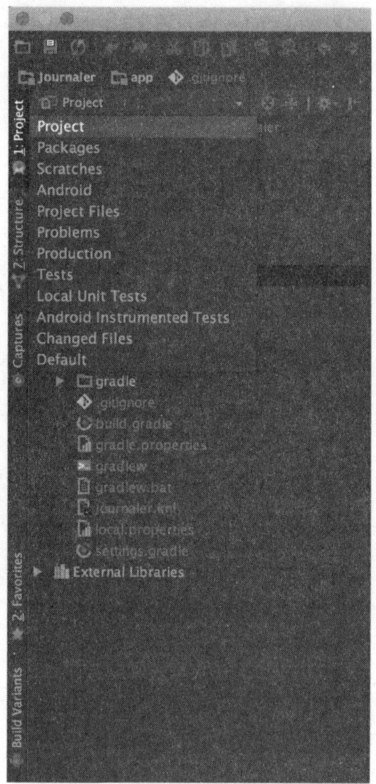

图 1.13

随后添加下列代码行：

```
.idea
.gradle
build/
gradle*
!gradle-plugins*
gradle-app.setting
!gradle-wrapper.jar
.gradletasknamecache
local.properties
gen
```

编辑位于 app 模块目录下的.gitignore 文件，如下所示。

```
*.class
.mtj.tmp/
```

```
*.jar
*.war
*.ear
hs_err_pid*
.idea/*
.DS_Store
.idea/shelf
/android.tests.dependencies
/confluence/target
/dependencies
/dist
/gh-pages
/ideaSDK
/android-studio/sdk
out
tmp
workspace.xml
*.versionsBackup
/idea/testData/debugger/tinyApp/classes*
/jps-plugin/testData/kannotator
ultimate/.DS_Store
ultimate/.idea/shelf
ultimate/dependencies
ultimate/ideaSDK
ultimate/out
ultimate/tmp
ultimate/workspace.xml
ultimate/*.versionsBackup
.idea/workspace.xml
.idea/tasks.xml
.idea/dataSources.ids
.idea/dataSources.xml
.idea/dataSources.local.xml
.idea/sqlDataSources.xml
.idea/dynamic.xml
.idea/uiDesigner.xml
.idea/gradle.xml
.idea/libraries
.idea/mongoSettings.xml
*.iws
/out/
.idea_modules/
```

```
atlassian-ide-plugin.xml
com_crashlytics_export_strings.xml
crashlytics.properties
crashlytics-build.properties
fabric.properties
target/
pom.xml.tag
pom.xml.releaseBackup
pom.xml.versionsBackup
pom.xml.next
release.properties
dependency-reduced-pom.xml
buildNumber.properties
.mvn/timing.properties
!/.mvn/wrapper/maven-wrapper.jar
samples/*
build/*
.gradle/*
!libs/*.jar
!Releases/*.jar
credentials*.gradle
gen
```

根据上述.gitignore 配置内容，随后即可执行确认和推送操作。在 macOS 操作系统上，可使用 Cmd+9 快捷键；在 Windows/Linux 操作系统上，可使用 Ctrl+9 快捷键（对应的快捷方式为 View | Tool Windows | Version Control）。随后，可展开未经版本控制的文件、选择这些文件并右击 Add to VCS，如图 1.14 所示。

图 1.14

接下来，按 Cmd+K 快捷键（在 Windows/Linux 操作系统上为 Ctrl+K 快捷键）、选中全部文件、输入 commit 消息，并从 Commit 下拉菜单中选择 Commit and Push。如果此时得到 Line Separators Warning，则可选择 Fix and Commit。此时将会显示 Push Commits 窗口。接下来，选中 Push Tags 复选框并选择 Current Branch，最后单击 Push 按钮。

1.9 设置 Gradle

Gradle 是一个构建系统。当然，用户可以不使用 Gradle 构建 Android 应用程序，但需要使用多个 SDK 工具，这也使得构建过程变得复杂起来。对此，我们需要使用 Gradle 和 Android Gradle 插件。

Gradle 获取所有的源文件并使用我们提到的工具进行处理。然后，它将所有内容打包到一个扩展名为.apk 的压缩文件中，则 APK 可以被解压。如果通过将其扩展名更改为 zip，进而对其进行重命名，则可以提取其中的内容。

每个构建系统都会使用自身的约定，其中较为重要的约定是将源代码和数据资源置于具有适当结构的目录中。

Gradle 是一个基于 JVM 的构建系统，因而可在 Java、Groovy、Kotlin 中编写脚本。另外，Gradle 还是一个基于插件的系统且易于扩展。一个较好的例子是谷歌的 Android 插件。读者可能注意到了项目中的 build.gradle 文件，此类文件均采用 Groovy 进行编写，因而所编写的任何 Groovy 代码均可被执行。下面将定义 Gradle 脚本以自动化一个构建过程。打开 settings.gradle 并查看下列内容：

```
include ":App"
```

该指令通知 Gradle 将构建一个名为 App 的模块，该 App 模块位于项目的 app 目录中。在项目的 root 中打开 build.gradle 文件，并添加下列代码行：

```
buildscript {
  repositories {
    jcenter()
    mavenCentral()
  }
  dependencies {
    classpath 'com.android.tools.build:gradle:2.3.3'
    classpath 'org.jetbrains.kotlin:kotlin-gradle-plugin:1.1.3'
  }
}
repositories {
  jcenter()
  mavenCentral()
}
```

这里，我们定义了构建脚本将从 JCenter 和 Maven Central 存储库解析其依赖关系。

同一存储库还将用于解析项目依赖关系。另外，还将添加主依赖关系并定位所包含的每个模块，具体如下：

- ❑ Android Gradle 插件。
- ❑ Kotlin Gradle 插件。

在更新了 build.gradle 主配置后，打开位于 App 模块目录下的 build.gradle 文件，并添加下列代码行：

```
apply plugin: "com.android.application"
apply plugin: "kotlin-android"
apply plugin: "kotlin-android-extensions"
android {
  compileSdkVersion 26
  buildToolsVersion "25.0.3"
  defaultConfig {
    applicationId "com.journaler"
    minSdkVersion 19
    targetSdkVersion 26
    versionCode 1
    versionName "1.0"
    testInstrumentationRunner
    "android.support.test.runner.AndroidJUnitRunner"
  }
  buildTypes {
    release {
      minifyEnabled false
      proguardFiles getDefaultProguardFile('proguardandroid.
      txt'), 'proguard-rules.pro'
    }
  }
  sourceSets {
    main.java.srcDirs += 'src/main/kotlin'
  }}
  repositories {
    jcenter()
    mavenCentral()
  }dependencies {
    compile "org.jetbrains.kotlin:kotlin-stdlib:1.1.3"
    compile 'com.android.support:design:26+'
    compile 'com.android.support:appcompat-v7:26+'}
```

上述配置可将 Kotlin 用作项目和 Gradle 脚本的开发语言，随后，代码定义了应用程序所需的最小和目标 SDK 版本——在当前示例中对应值分别为 19 和 26。需要注意的是，在默认配置部分中，我们还设置了应用程序 ID 和版本参数。依赖关系部分则设置了 Kotlin 自身和一些 Android UI 组件的依赖项（稍后将对此加以讨论）。

1.10 目录结构

Android Studio 包含了构建应用程序所需的一切内容，如源代码和数据资源。其中，所有的目录均是创建项目时的配置向导所生成的。当查看目录时，可打开 IDE 左侧的 Project 窗口（选择 View | ToolWindows | Project），如图 1.15 所示。

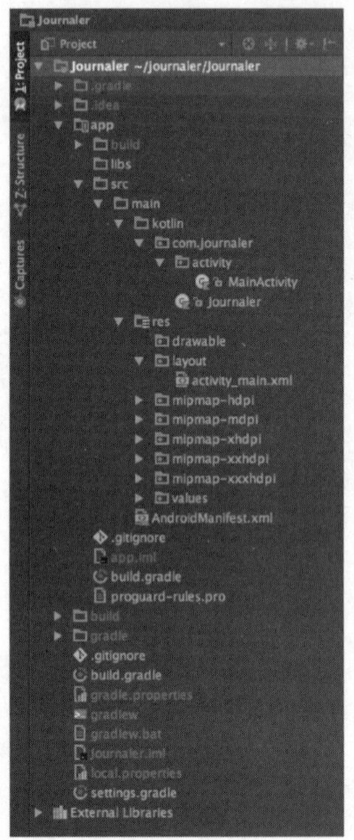

图 1.15

项目模块表示源文件、资产和构建设置的集合，这些设置将项目划分为独立的功能部件。其中，最小的 modules 数量为 1；而项目中包含的最大 modules 数量则无限制。另外，还可分别对 modules 进行构建、测试和调试。当前，我们所定义的 Journaler 项目中仅包含了一个名为 app 的模块。

下面尝试构建一个新的模块。对此，访问 File | New | New Module，如图 1.16 所示。

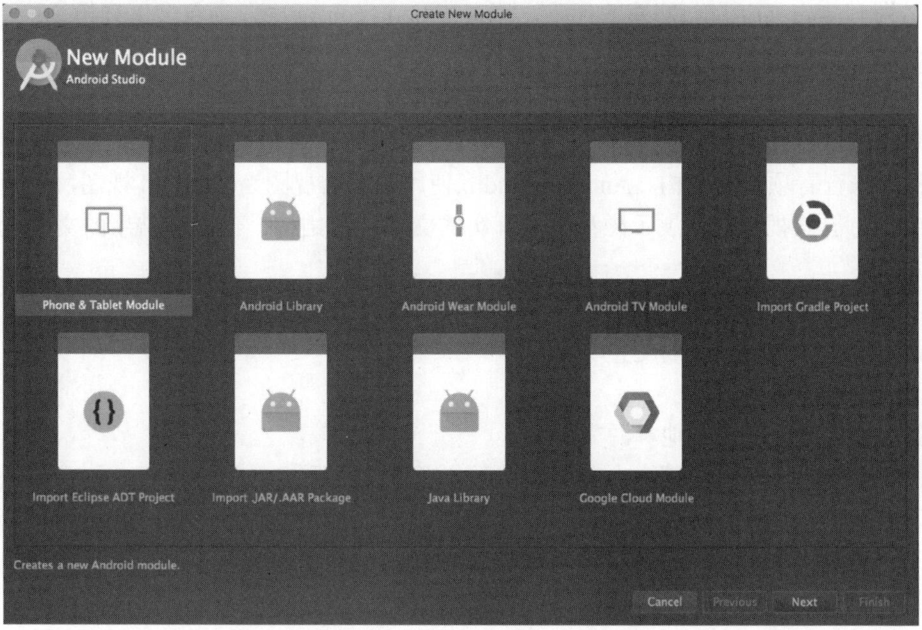

图 1.16

此处可能会生成下列模块：

- Android Application Module 表示为应用程序源代码、资源和设置的容器。相应地，默认的模块名为 app，这与当前示例中的名称保持一致。
- Phone & Tablet Module。
- Android Wear Module。
- Glass Module。
- Android TV module。
- Library 模块表示可复用的代码容器——库。该模块可用作其他应用程序模块中的依赖关系，或者被导入其他项目中。当构建该模块时，将包含一个 AAR 扩展名，即 Android 存档而非 APK 扩展。

Create New Module 模块提供了下列选项：

- Android Library：Android 项目中的全部类型均被支持。该库的构建结果为 Android Archiver（AAR）。
- Java Library：仅支持 Java。该库的构建结果为 Java Archiver（JAR）。
- Google Cloud Module：针对 Google Cloud 后端代码定义了一个容器。

需要着重理解的是，Gradle 将 modules 引为一个独立的项目。如果应用程序代码依赖于名为 Logger 的 Android 库的代码，那么在 build.config 中，须包含下列指令：

```
dependencies {
  compile project(':logger')
}
```

接下来查看项目的结构。Android Studio 使用的默认视图为 Android 视图，进而显示项目文件；但该视图并未展示磁盘上的真实的文件层次结构，一些不常用的文件或目录将被隐藏。

Android 视图显示了下列内容：

- 所有与构建相关的配置文件。
- 所有清单文件。
- 独立分组中的其他资源文件。

在每个应用程序中，模块内容通过以下分组予以表示：

- 清单文件和 AndroidManifest.xml 文件。
- 应用程序和测试的 Java 和 Kotlin 源代码。
- res 和 Android UI 资源。
- 当查看项目的实际文件结构时，可执行 Project view 命令。对此，可单击 Android view，并从下拉菜单中选择 Project。

据此，我们可查看到更多的文件和目录，其中较为重要的内容包括：

- module-name/：表示为模块的名称。
- build/：包含了构建的输出结果。
- libs/：包含了私有库。
- src/：涵盖了下列子目录中模块的代码和源文件：
  - main：包含了 main 源文件，即所有构建变化版本的源代码和资源（稍后将对构建变化版本加以讨论）。
  - AndroidManifest.xml：该文件定义了应用程序及其每个组件的性质。
  - java：包含了 Java 源代码。
  - kotlin：包含了 Kotlin 源代码。
  - jni：包含了基于 Java 本地接口（Java Native Interface，JNI）的本地代码。
  - gen：包含了 Android Studio 生成的 Java 文件。
  - res：包含了应用程序资源，如 drawable 文件、布局文件、字符串等。
  - assets：包含了编译为.apk 文件的文件（无改动）。

- test：包含了测试源代码。
- build.gradle：表示为项目级别的构建配置内容。

选择 File | Project Structure 可改变项目设置，如图 1.17 所示。

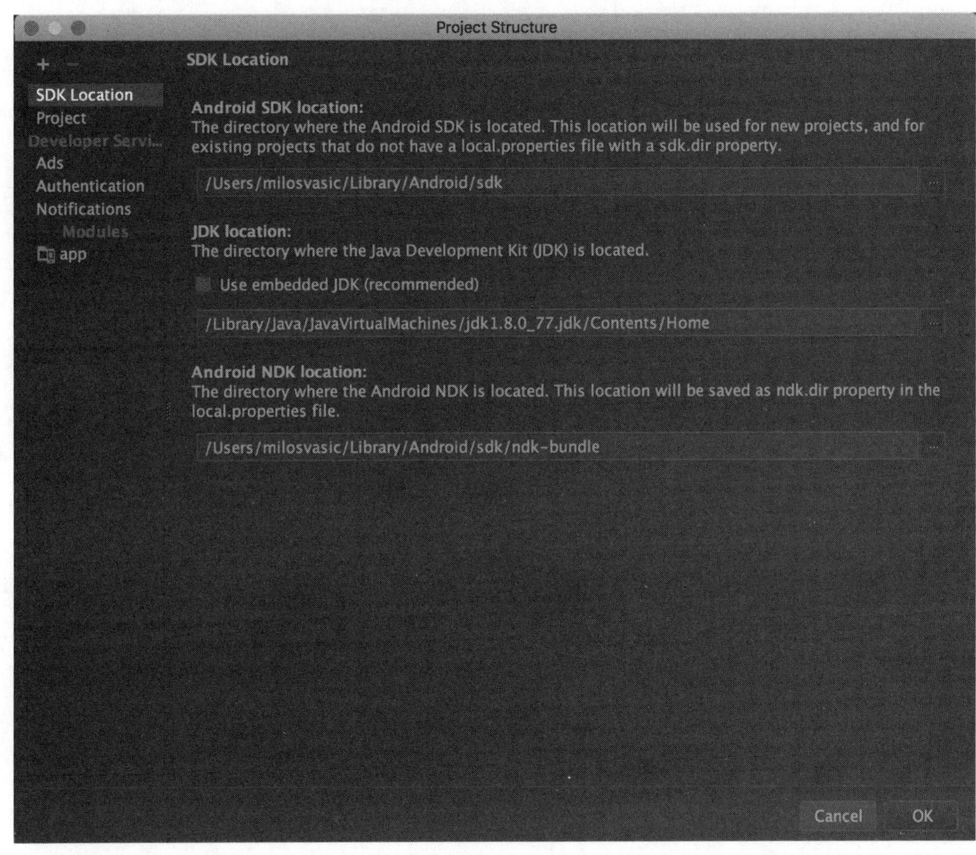

图 1.17

其中包含了以下各项内容：

- SDK Location：表示为项目所使用的 Android SDK、JDK 和 Android NDK 的位置。
- Project：用于设置 Gradle 和 Android Gradle 插件版本。
- Modules：用于编辑特定模块的构建配置内容。

相应地，Modules 部分可分为以下选项卡：

- Properties：用于构建模块时设置 SDK 和构建工具的版本。
- Signing：用于设置 APK 签名的证书。
- Flavors：用于定义模块的风格。

- ❑ Build Types：用于定义模块的构建类型。
- ❑ Dependencies：用于设置模块所需的依赖关系。

对应结果如图 1.18 所示。

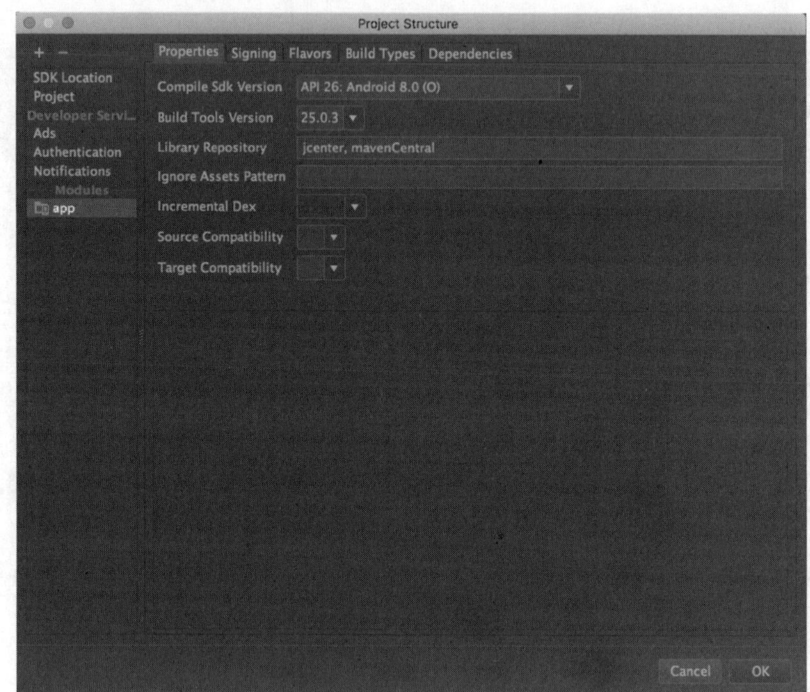

图 1.18

1.11 定义构建类型和风格

本节讨论项目的重要阶段，即定义应用程序构建的 Build Variants。这里，Build Variants 代表了 Android 应用程序的唯一版本。

这里，唯一性是指覆写了某些应用程序属性和资源。

另外，每个 Build Variants 将在模块级别上进行配置。

下面将尝试扩展 build.gradle，并将下列代码置于 build.gradle 文件的 android 部分中：

```
android {
  ...
  buildTypes {
```

```
    debug {
      applicationIdSuffix ".dev"
    }
    staging {
      debuggable true
      applicationIdSuffix ".sta"
    }
    preproduction {
      applicationIdSuffix ".pre"
    }
      release {}
    }
   ...
}
```

此处针对应用程序分别定义了下列 buildTypes：release、staging、preproduction 和 debug。相应地，产品风格的创建方式与 buildTypes 类似，可将其添加至 productFlavors 中，并对所需的设置项进行配置，如下所示。

```
android {
  ...
  defaultConfig {...}
  buildTypes {...}
  productFlavors {
    demo {
      applicationIdSuffix ".demo"
      versionNameSuffix "-demo"
    }
    complete {
      applicationIdSuffix ".complete"
      versionNameSuffix "-complete"
    }
    special {
      applicationIdSuffix ".special"
      versionNameSuffix "-special"
    }
  }
}
```

在创建和配置了 productFlavors 之后，可单击通知栏中的 Sync Now 按钮。

此时，用户需要稍等片刻以完成处理过程。Build Variants 的名称由<product-flavor><Build-Type>规则构成，相关示例如下：

```
demoDebug
demoRelease
completeDebug
completeRelease
```

当然，用户可根据需要修改 Build Variant。对此，可访问 Build 并选择 Build Variant，同时从下拉列表中选取 completeDebug，如图 1.19 所示。

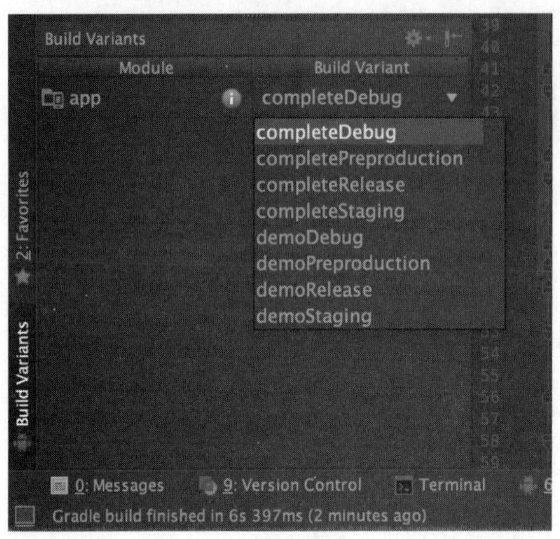

图 1.19

Main/source 集合在应用程序 Build Variants 间所共享。如果需要创建一个新的源集合，则需要针对特定的构建类型、构建风格及其组合予以实施。

类似于 Main/Source 集合，所有的源集合文件和目录都需要采用特定方式加以组织。针对调试构建类型的 Kotlin 类文件须位于 src/debug/kotlin/directory 中。

针对文件的组织方式，可打开终端窗口（View | ToolWindows | Terminal）并执行下列命令行：

```
./gradlew sourceSets
```

用户可执行查看输出结果，该结果易于理解且具有自解释性。另外，Android Studio 并不能创建 sourceSets，这需要用户亲自完成。

必要时，还可以更改 Gradle 使用 sourceSets 块查找源集合的位置。下面将更新所期望的源代码路径：

```
android {
  ...
```

```
  sourceSets {
    main {
      java.srcDirs = [
                'src/main/kotlin',
                'src/common/kotlin',
                'src/debug/kotlin',
                'src/release/kotlin',
                'src/staging/kotlin',
                'src/preproduction/kotlin',
                'src/debug/java',
                'src/release/java',
                'src/staging/java',
                'src/preproduction/java',
                'src/androidTest/java',
                'src/androidTest/kotlin'
      ]
      ...
    }
  }
}
```

对于希望采用特定配置打包的代码和资源，可将其存储至 sourceSets 目录中。下面给出了采用 demoDebug 构建变化版本进行构建的示例，该构建变化版本演示了 demo 产品风格和 debug 构建类型。在 Gradle 中，对应的优先顺序如下：

```
src/demoDebug/ (build variant source set)
src/debug/ (build type source set)
src/demo/ (product flavor source set)
src/main/ (main source set)
```

这是 Gradle 在构建过程中使用的优先级顺序，且需要在应用以下构建规则时予以考虑：

- 编译 java/和 kotlin/目录中的源代码时。
- 将清单文件一并合并至独立的清单文件时。
- 合并 values/目录中的文件时。
- 合并 res/和 asset/目录中的资源时。

相应地，库模块依赖项中包含的资源和清单的优先级最低。

1.12　附　加　库

前述内容配置了构建类型和风格，本节将引入一些第三方库，如 Retrofit、OkHttp 和

Gson，具体如下：

- Retrofit 是 Square, Inc 推出的、针对 Android 和 Java 的、类型安全的 HTTP 客户端。就简洁性和性能而言，Retrofit 是较为流行的 Android 客户端库之一。
- OkHttp 是一种默认情况下高效的 HTTP 客户端——HTTP/2 允许对同一主机的所有请求共享一个套接字。
- Gson 则是一个 Java 库，用于将 Java 对象转换为 JSON 表达形式。另外，Gson 还可将 JSON 字符串转换为对应的 Java 对象。Gson 可以处理任意 Java 对象，包括尚未包含源代码的已有对象。

相应地，存在多个开源项目可将 Java 对象转换为 JSON。稍后，我们将引入 Kotson，并针对 Kotlin 提供 Gson 绑定。

下面利用 Retrofit 和 Gson 依赖关系扩展 build.gradle，如下所示。

```
dependencies {
  ...
  compile 'com.google.code.gson:gson:2.8.0'
  compile 'com.squareup.retrofit2:retrofit:2.2.0'
  compile 'com.squareup.retrofit2:converter-gson:2.0.2'
  compile 'com.squareup.okhttp3:okhttp:3.6.0'
  compile 'com.squareup.okhttp3:logging-interceptor:3.6.0'
  ...
}
```

在更新了 Gradle 配置后，需要再次对其进行同步。

1.13　Android Manifest

每个应用程序都需要设置一个 AndroidManifest.xml 文件，该文件位于 root 目录中。在每个模块中，该文件包含了应用程序针对 Android 系统的必要信息。manifest 文件负责定义下列内容：

- 命名应用程序的数据包。
- 描述应用程序的组件——活动（屏幕）、服务、广播接收者（消息）以及内容提供者（数据库访问）。
- 应用程序授权，进而可访问受保护的 Android API。
- 其他应用程序的授权，进而与应用程序组件进行交互，例如内容提供者。

下列代码片段显示了 manifest 文件及其所包含元素的一般结构：

```
<?xml version="1.0" encoding="utf-8"?>
<manifest>
 <uses-permission />
 <permission />
 <permission-tree />
 <permission-group />
 <instrumentation />
 <uses-sdk />
 <uses-configuration />
 <uses-feature />
 <supports-screens />
 <compatible-screens />
 <supports-gl-texture />

 <application>
   <activity>
     <intent-filter>
       <action />
         <category />
           <data />
       </intent-filter>
       <meta-data />
   </activity>

   <activity-alias>
     <intent-filter> . . . </intent-filter>
     <meta-data />
   </activity-alias>

   <service>
     <intent-filter> . . . </intent-filter>
     <meta-data/>
   </service>

   <receiver>
     <intent-filter> . . . </intent-filter>
     <meta-data />
   </receiver>
   <provider>
     <grant-uri-permission />
     <meta-data />
     <path-permission />
```

```
    </provider>

    <uses-library />
  </application>
</manifest>
```

1.14　主应用程序类

每个 Android 应用程序均定义了其 Application 主类。Android 中的 Application 类表示为 Android 应用程序中的基类，并涵盖了所有其他组件，如 activities 和 services。在创建应用程序/包时，Application 类（或者其子类）须在其他类之前予以实例化。

下面将针对 Journaler 定义 Application 类。对此，可定位主源目录；如果不存在 Kotlin 源目录，则创建该目录。随后，创建数据包 com 以及子包 journaler。对此，可右击 Kotlin 目录并选择 New | Package。一旦创建了包结构，可右击 journaler 包并选择 New | KotlinFile/Class，将其命名为 Journaler。此时将生成 Journaler.kt 文件。

每个 Application 类需要扩展 Android Application 类，如下所示。

```
package com.journaler

import android.app.Application
import android.content.Context

class Journaler : Application() {

  companion object {
    var ctx: Context? = null
  }

  override fun onCreate() {
    super.onCreate()
    ctx = applicationContext
  }

}
```

当前，Application 主类提供了应用程序上下文的静态访问，稍后将对上下文这一概念加以解释。然而，Android 并不会使用该类，直至该类在清单文件中被提及。对此，打开 app 模块 android manifest 并添加下列代码块：

```
<manifest xmlns:android="http://schemas.android.com/apk/
res/android" package="com.journaler">

<application
  android:name=".Journaler"
  android:allowBackup="false"
  android:icon="@mipmap/ic_launcher"
  android:label="@string/app_name"
  android:roundIcon="@mipmap/ic_launcher_round"
  android:supportsRtl="true"
  android:theme="@style/AppTheme">

</application>
</manifest>
```

通过 android:name=".Journaler"，将通知 Android 使用哪一个类。

1.15　第一个屏幕画面

前述内容创建了一个未包含任何屏幕画面的应用程序，本节将讨论如何生成一个屏幕画面。对此，创建一个名为 activity 的新数据包，并于其中定义所有的屏幕类，同时创建名为 MainActivity.kt 的第一个 Activity 类。下列代码显示了一个稍微简单的类：

```
package com.journaler.activity

import android.os.Bundle
import android.os.PersistableBundle
import android.support.v7.app.AppCompatActivity
import com.journaler.R

class MainActivity : AppCompatActivity() {
  override fun onCreate(savedInstanceState: Bundle?,
  persistentState: PersistableBundle?) {
    super.onCreate(savedInstanceState, persistentState)
    setContentView(R.layout.activity_main)
  }
}
```

稍后将解释每行代码的具体含义。当前，需要注意的是，setContentView(R.layout.activity_main)为屏幕分配 UI 资源，activity_main 则是定义它的 XML 名称。下面将尝试

对此予以构建。访问 main 目录下的 res 目录，如果尚不存在 layout 文件夹，则对此加以创建，并于随后创建一个名为 activity_main 的新布局，即右击 layout 目录并选择 New | Layout 资源文件，将 activity_main 指定为它的名称，并将 LinearLayout 指定为它的根元素。该文件内容如下：

```
  <?xml version="1.0" encoding="utf-8"?>
  <LinearLayout xmlns:android="http://schemas.android.com/
    apk/res/android"
      android:orientation="vertical"
      android:layout_width="match_parent"
      android:layout_height="match_parent">

</LinearLayout>
```

在运行应用程序之前，还需要注意以下一点：通知清单文件与屏幕相关的信息。因此，打开 main manifest 文件并加入下列代码片段：

```
<application ... >
  <activity
    android:name=".activity.MainActivity"
    android:configChanges="orientation"
    android:screenOrientation="portrait">
    <intent-filter>
      <action android:name="android.intent.action.MAIN" />
      <category android:name="android.intent.category.LAUNCHER" />
    </intent-filter>
  </activity>
</application>
```

稍后将对其中的每个属性予以解释。当前，应用程序已处于运行就绪状态。

1.16 本章小结

本章介绍了 Android 的一些基础知识，以及少量的 Kotlin 方面的内容。此外，本章还配置了 Android 的工作环境，并创建了第一个应用程序画面。

第 2 章将深入讨论 Android 方面的内容，即如何构建应用程序，并自定义不同的版本。除此之外，我们还将探讨运行应用程序的各种方式。

第 2 章　构建和运行应用程序

截至目前，我们已经构建了包含单一屏幕画面的 Android 项目。第 1 章中介绍了如何设置相应的工作环境、如何使用简单的 Android 工具，并定义了相关风格和构建类型。本章将在设备或模拟器上运行应用程序，同时尝试使用各种构建类型和风格组合。

本章主要涉及以下主题：

- 在模拟器和/或真实设备上运行应用程序。
- 日志简介。
- Gradle 工具。

2.1　运行第一个 Android 应用程序

在第 1 章中，我们生成了第一个屏幕画面，并针对应用程序自身定义了一组规范。为了确保一切正常，还需要构建和运行应用程序。这里将运行 completeDebug 构建变化版本。关于如何切换至该构建变化版本中，我们将在操作过程中予以提示。

打开 Android Studio 和 Journaler 项目；随后单击 Android Studio 窗口左侧的 Build Variants 面板，进而打开 Build Variants 面板；或者通过选择 View |Tool Windows | Build Variants 命令。此时将显示 Build Variants 面板，并从下拉列表中选取 completeDebug，如图 2.1 所示。

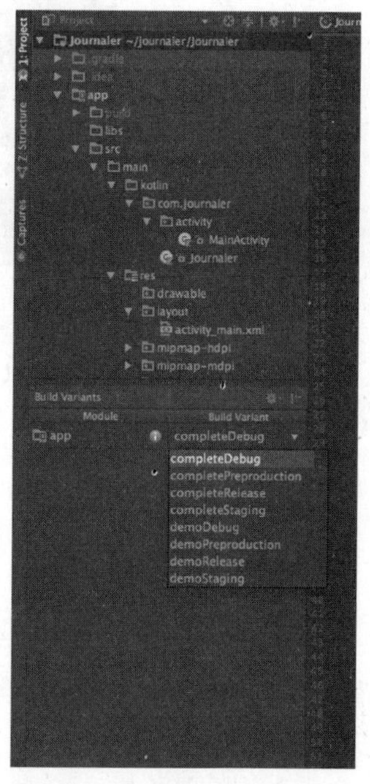

图 2.1

我们将使用图 4.1 中的 Build Variant 作为主构建变体；对于产品构建，我们将采用 completeDebug 构建变化版本。在从下拉列表中选择了 Build Variant 后，对于 Gradle 来说，将花费些许时间构建所选择的变化版本。

下面将尝试运行应用程序。首先，应用程序将在模拟器中运行，随后则在真实设备上运行。对此，打开 AVD Manager 启动模拟器实例，此处可单击 AVD Manager 图标，这也是一种较快的开启方式。如果双击 AVD 实例，这将花费一些时间直至模拟器处于就绪状态。模拟器执行 Android 系统引导，然后加载一个默认的应用程序启动器。

当前，模拟器处于启动状态并可运行应用程序。相应地，可单击 Run 图标或者选择 Run | Run 'app'命令。

注意：

macOS 上的快捷方式为 Ctrl+R 快捷键。

当应用程序处于运行状态时，将显示一个 Select Deployment Target 对话框。其中，应用程序可在多个实例上运行，用户可选取其中的一个，如图 2.2 所示。

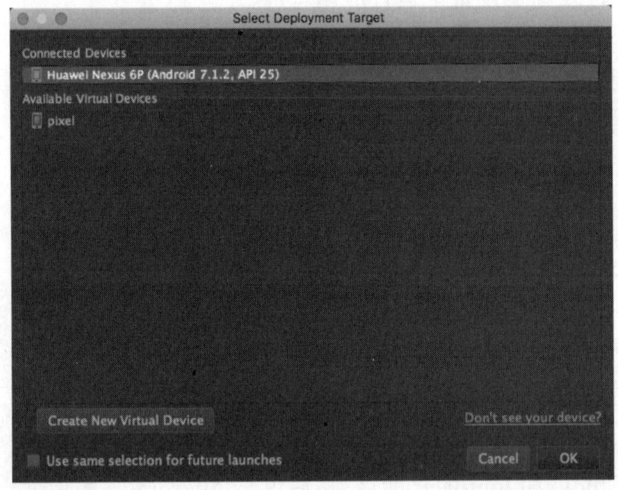

图 2.2

选择部署目标并单击 OK 按钮。当希望记起选项结果时，可选中 Use same selection for future launches 复选框。运行应用程序将会占用少许时间。

2.2　Logcat

Logcat 是开发过程中较为重要的一个环节，旨在显示设备中的全部日志消息。具体来说。Logcat 将显示连接自模拟器和真实设备上的日志内容。Android 提供了多种日志消息级别，如下所示。

- 断言（Assert）。
- 调试（Debug）。

- 错误（Error）。
- 信息（Info）。
- 显示详细内容（Verbose）。
- 警告（Warning）。

通过上述日志级别（例如仅需要查看错误时——应用程序崩溃栈跟踪）、日志标签（稍后将对此加以解释）、关键字、正则表达式或应用程序包可对日志消息进行过滤。

选择 Android Studio | Preferences 命令，在搜索框内输入 Logcat，随后将显示一些偏好设置，如图 2.3 所示。

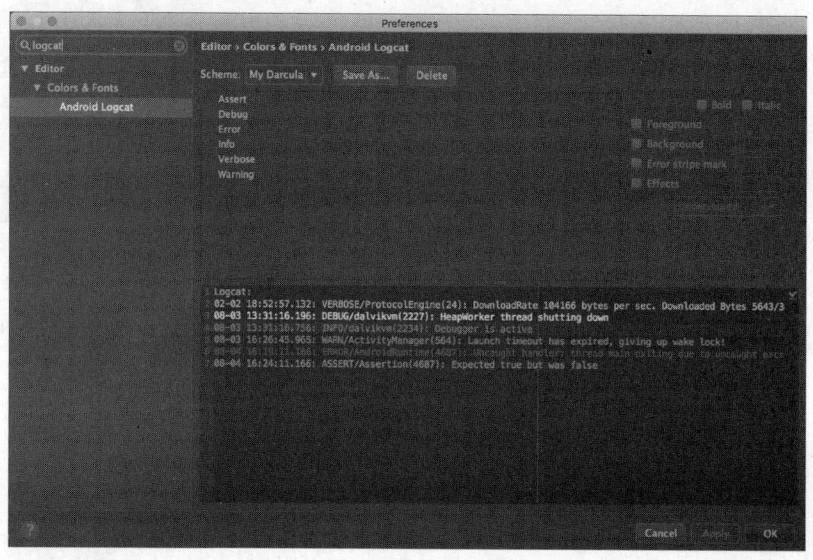

图 2.3

当对色彩进行编辑时，需要保存当前颜色主题的副本。在主题下拉列表中选取相应的主题，并单击 Save As 按钮。此处需要选取相应的名称并予以确认，如图 2.4 所示。

从列表中选择 Assert 并取消选中 Use inherited attributes 复选框以覆盖当前颜色。这里，应确保选中 Foreground 复选框，并单击位于复选框右侧的 Color，进而针对日志文本选择新的颜色，如图 2.5 所示。

对于 Assert 级别，可通过手动方式输入十六进制代码 FF6B68。就可读性来说，此处建议使用下列颜色：

- Assert：#FF6B68。
- Verbose：#BBBBBB。
- Debug：#F4F4F4。

- Information：#6D82E3。
- Warning：#E57E15。
- Error：#FF1A11。

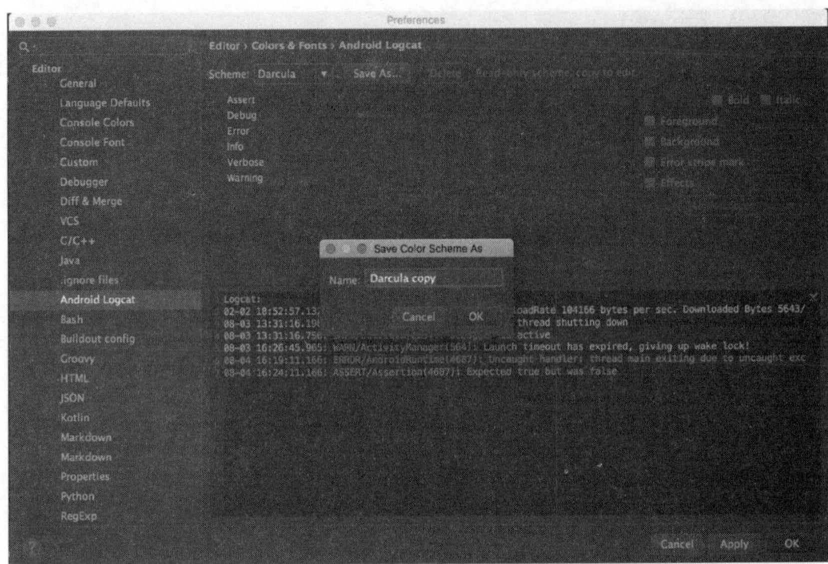

图 2.4

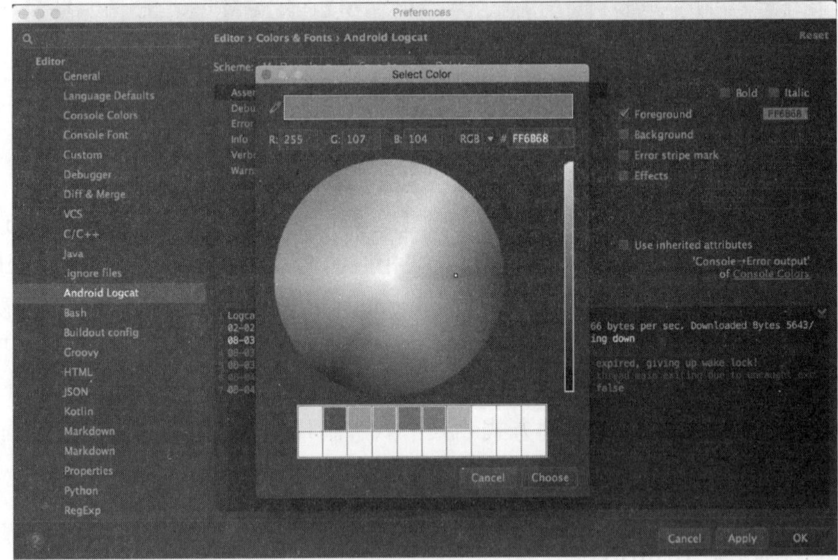

图 2.5

当应用更改结果时，则单击 Apply 按钮，然后再单击 OK 按钮。

打开 Android Monitor（View | Tool Windows | Android Monitor 命令）并查看 Logcat 面板中输出的消息。此时，针对不同的日志级别，将通过不同的颜色加以显示，如图 2.6 所示。

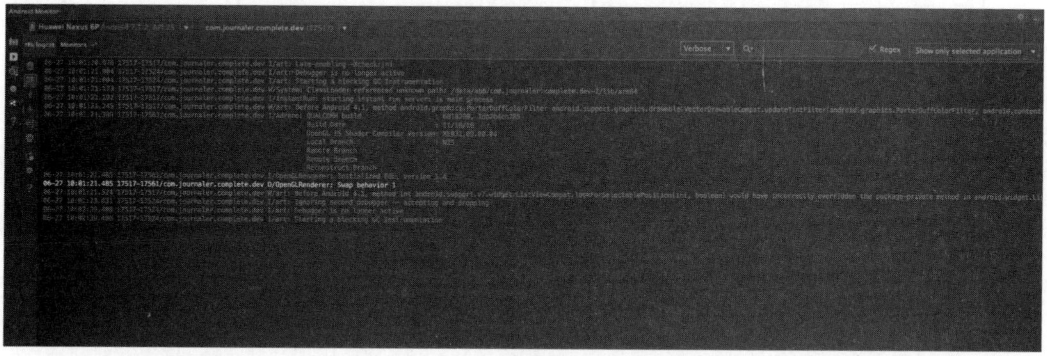

图 2.6

下面将尝试定义自己的日志消息，与此同时，我们也能够进一步了解如何与 Android 生命周期协同工作。相应地，针对所创建的 Application 类和屏幕（活动）的每个生命周期事件，我们将设置相应的日志消息。

打开 Application 类所属的 Journaler.kt 文件并调整代码，如下所示。

```
class Journaler : Application() {

 companion object {
  val tag = "Journaler"
  var ctx: Context? = null
 }

 override fun onCreate() {
  super.onCreate()
  ctx = applicationContext
  Log.v(tag, "[ ON CREATE ]")
 }

 override fun onLowMemory() {
  super.onLowMemory()
  Log.w(tag, "[ ON LOW MEMORY ]")
 }
```

```
  override fun onTrimMemory(level: Int) {
    super.onTrimMemory(level)
    Log.d(tag, "[ ON TRIM MEMORY ]: $level")
  }
}
```

这里，我们引入了一些重要的改变，代码中覆写了 onCreate()应用的主要生命周期事件。除此之外，还覆写了两个附加方法。其中，onLowMemory()方法在出现较为严重的内存问题时将被触发（正常的运行处理进程应清理内存使用）；当内存被清理时，将触发 onTrimMemory()方法。

对于应用程序中的日志事件，可采用包含静态方法（分别对应于相应的日志级别）的 Log 类。据此，我们可公开下列方法：

- 对于 Verbose 级别，如下所示。

```
v(String tag, String msg)
v(String tag, String msg, Throwable tr)
```

- 对于 Debug 级别，如下所示。

```
d(String tag, String msg)
d(String tag, String msg, Throwable tr)
```

- 对于 Information 级别，如下所示。

```
i(String tag, String msg)
i(String tag, String msg, Throwable tr)
```

- 对于 Warning 级别，如下所示。

```
w(String tag, String msg)
w(String tag, String msg, Throwable tr)
```

- 对于 Error 级别，如下所示。

```
e(String tag, String msg)
e(String tag, String msg, Throwable tr)
```

上述各方法分别接收下列参数：

- Tag：该参数用于确认日志消息的来源。
- message：该参数表示为日志消息。
- throwable：该参数表示为日志异常。

除了上述日志方法之外，我们还可使用一些附加的方法，如下所示。

- wtf(String tag, String msg)方法。
- wtf(String tag, Throwable tr)方法。
- wtf(String tag, String msg, Throwable tr)方法。

稍后还将对 Log 类进行更多尝试。下列代码展示了 MainActivity 类更新后的内容。

```
class MainActivity : AppCompatActivity() {
 private val tag = Journaler.tag

 override fun onCreate(
  savedInstanceState: Bundle?,
  persistentState: PersistableBundle?
 ){
    super.onCreate(savedInstanceState, persistentState)
    setContentView(R.layout.activity_main)
    Log.v(tag, "[ ON CREATE ]")
  }

 override fun onPostCreate(savedInstanceState: Bundle?) {
  super.onPostCreate(savedInstanceState)
  Log.v(tag, "[ ON POST CREATE ]")
 }

 override fun onRestart() {
  super.onRestart()
  Log.v(tag, "[ ON RESTART ]")
 }

 override fun onStart() {
  super.onStart()
  Log.v(tag, "[ ON START ]")
 }

 override fun onResume() {
  super.onResume()
  Log.v(tag, "[ ON RESUME ]")
 }

 override fun onPostResume() {
  super.onPostResume()
```

```
    Log.v(tag, "[ ON POST RESUME ]")
  }

  override fun onPause() {
    super.onPause()
    Log.v(tag, "[ ON PAUSE ]")
  }

  override fun onStop() {
    super.onStop()
    Log.v(tag, "[ ON STOP ]")
  }

  override fun onDestroy() {
    super.onDestroy()
    Log.v(tag, "[ ON DESTROY ]")
  }
}
```

此处根据活动生命周期内的执行顺序覆写了所有的重要方法。针对每个事件，将输出相应的日志消息。下面将解释生命周期和每个重要事件的含义。

图 2.7 中显示了 Android 开发者网站中提供的官方示意图,并解释了活动的生命周期。

读者可访问 https://developer.android.com/images/activity_lifecycle.png 以进一步查看图 2.7，其中的方法具体如下：

- onCreate()：当首次创建活动时将执行该方法。在该方法中，通常会执行 UI 主元素的初始化操作。
- onRestart()：活动在某一时刻终止并随后恢复时将执行该方法。例如关闭手机屏幕（锁屏），并于随后再次解屏。
- onStart()：当屏幕对应用程序用户可见时执行该方法。
- onResume()：当用户开始与当前活动进行交互时执行该方法。
- onPause()：在继续之前的活动之前，该方法将在当前活动上执行。其中可存储再次恢复时所需的全部信息。如果存在未保存的更改内容，可将其保存于此。
- onStop()：当活动对于应用程序用户不再可见时执行该方法。
- onDestroy()：某个活动在由 Android 销毁之前执行该方法。这种情况可能会存在，例如，某人执行了 Activity 类的 finish()方法。当希望了解当前活动是否在特定的时间点结束时，Android 提供了一个检测方法 isFinishing()。如果对应活动结束，

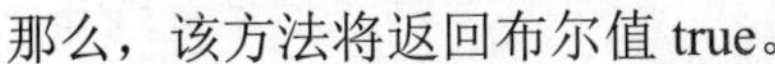
那么，该方法将返回布尔值 true。

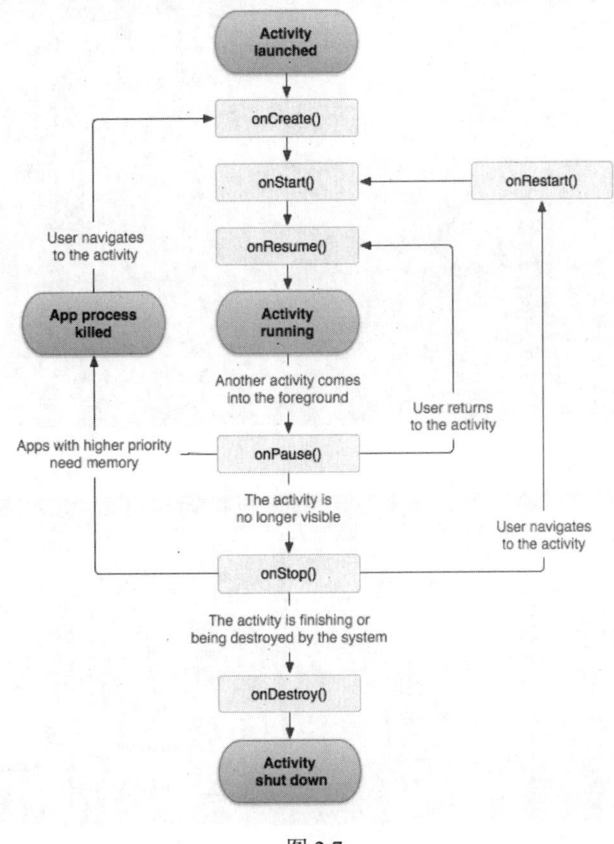

图 2.7

接下来将编写相关代码并生成相应的日志消息。对此，我们将执行两个用例，并查看 Logcat 输出的日志消息。

打开 Android Monitor，从下拉列表中选择设备实例（模拟器或真实设备）；随后从下一个下拉列表中选择 Journaler 应用程序包。图 2.8 显示了 Logcat 的输出结果。

其中可以看到源代码中置入的日志消息。

接下来查看在应用程序交互过程中，进入 onCreate()方法和 onDestroy()方法的次数。对此，在搜索栏中输入 on create，如图 2.9 所示。考查内容变化后将会发现，此时仅存在一项内容，而非期望中的两项（分别对应于 Application 主类和主活动），稍后将对此加以解释。

图 2.8

图 2.9

相应地，输出结果中包含了以下内容：

- ❑ 06-27：表示为事件产生的日期。
- ❑ 11:37:59.914：表示为事件产生的时间。
- ❑ 6713-6713/?：表示为包的进程和线程标识符。如果应用程序只有一个线程，则进程标识符和线程标识符是相同的。
- ❑ V/Journaler：表示为日志级别和标签。
- ❑ [ON CREATE]：表示为日志消息。

如果将过滤器调整为 on destroy，对应的内容变化如下所示。

```
06-27 11:38:07.317 6713-6713/com.journaler.complete.dev V/Journaler: [ ON DESTROY ]
```

其中包含了不同的日期、时间和 pid/tid 值。

在下拉列表中，将过滤机制从 Verbose 调整为 Warn，并维持过滤器值。此时将会发现，Logcat 中包含了空内容，其原因在于，此处不存在包含 on destroy 消息文本的警告消息。移除过滤器文本并返回 Verbose 级别。

运行应用程序并在一行中锁定/解锁屏幕。随后，关闭并杀死 Journaler 应用程序，Logcat 的输出结果如图 2.10 所示。

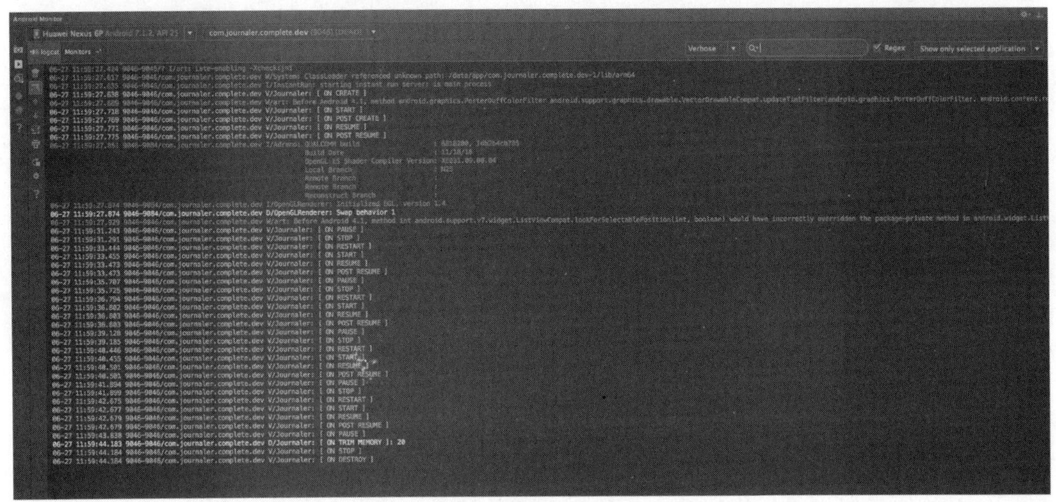

图 2.10

可以看到，当前进入了“暂停和恢复”生命周期状态。最后，我们将杀死当前应用程序，同时引发 onDestroy()事件，这可在 Logcat 中看到。

除此之外，还可在终端中使用 Logcat。对此，打开终端并输出下列命令行：

```
adb logcat
```

2.3　使用 Gradle 构建工具

在开发过程中，常需要创建不同的构建版本或与运行测试程序。必要时，此类测试可仅针对特定的构建变化版本加以执行（或者全部版本）。

下面将讨论某些较为常见的 Gradle 用例，并首先介绍清除和构建机制。

如前所述，Journaler 应用程序涵盖了下列已定义的构建类型：

- ❑ 调试。

- 发布。
- 暂存。
- 生产前。

另外，下列构建风格也定义于 Journaler 应用程序中：

- 演示版本。
- 完整版本。
- 特定版本。

打开终端，当移除之前所创建的所有内容，以及临时构建的衍生结果时，可执行下列命令：

```
./gradlew clean
```

清除操作可能会花费一些时间，随后可执行以下命令：

```
./gradlew assemble
```

这将整合全部内容——应用程序中所有的构建变化版本。想象一下，如果我们处理的是一个非常大的项目，时间方面往往会带来很大的影响。针对于此，可隔离构建命令。当仅创建调试构建类型时，可执行下列命令行：

```
./gradlew assembleDebug
```

与之前的示例相比，执行速度将得到显著的提升。这将针对调试构建类型创建全部风格。为了更加有效，可通知 Gradle 仅关注调试构建类型的完整构建风格。对此，可执行下列命令：

```
./gradlew assembleCompleteDebug
```

相应地，执行速度也将显著提升。这里，我们讨论了多种重要的 Gradle 命令。

当执行全部单元测试时，可执行下列命令：

```
./gradlew test
```

如果希望针对特定的构建变化版本执行单元测试，可执行下列命令：

```
./gradlew testCompleteDebug
```

在 Android 中，可在真实的设备实例或模拟器上运行测试。通常，这一类测试可访问某些 Android 组件。当执行这一类测试时，可使用下列命令：

```
./gradlew connectedCompleteDebug
```

在本书的后续章节中，还将对 Android 应用程序进行更多的测试。

2.4　调试应用程序

前述内容讨论了如何生成重要的应用程序消息日志。在开发过程中，在分析应用程序行为或研究 bug 时，仅记录日志消息有时是远远不够的。

更为重要的是，在 Android 真实设备或模拟器上，应用程序代码应可在执行过程中进行调试。对此，可打开 Application 主类，并在记录 onCreate()方法的代码行上设置断点，如图 2.11 所示。

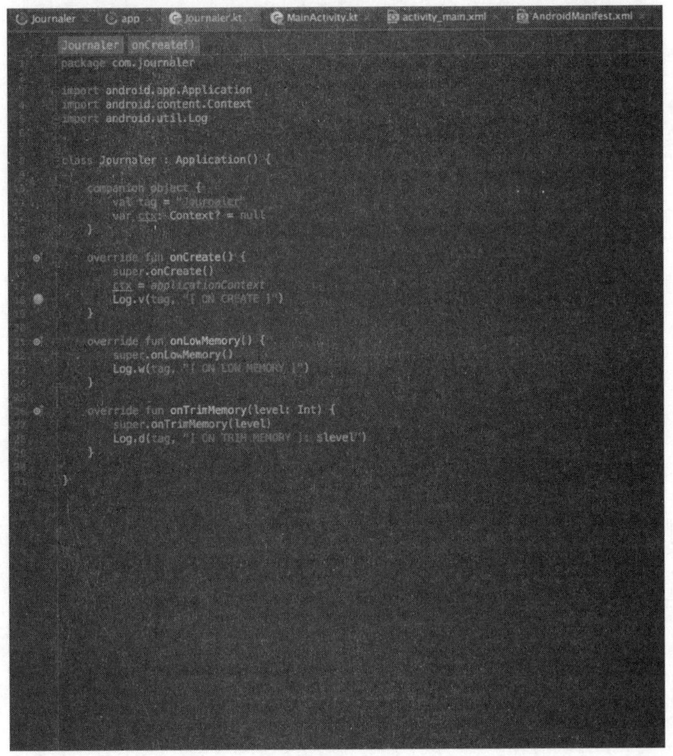

图 2.11

可以看到，当前在第 18 行上设置了断点。稍后，我们还将尝试添加更多的断点，下面将其添加至主（唯一）活动中。在图 2.12 中，我们在执行日志记录的相关行上为每个生命周期事件设置一个断点。

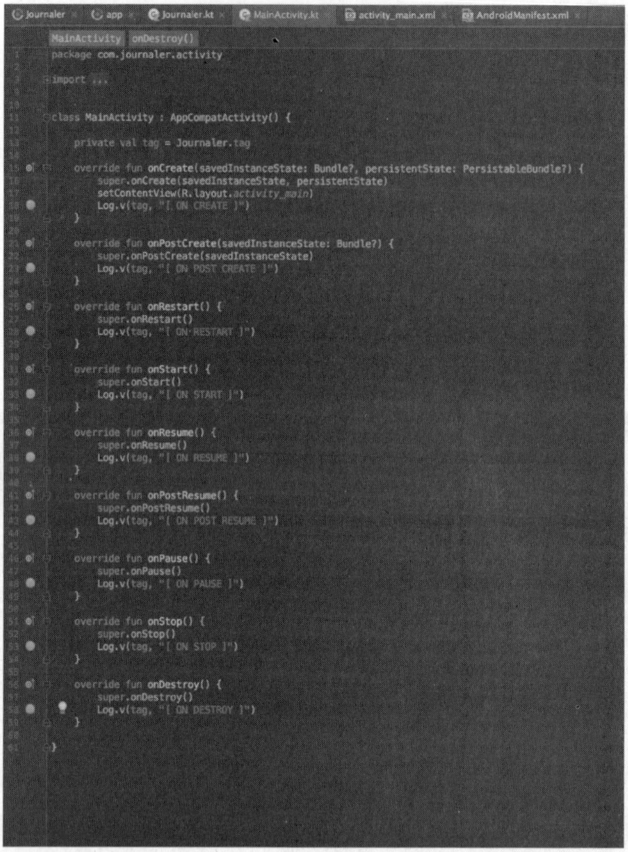

```
1  package com.journaler.activity
2
3  import ...
10
11 class MainActivity : AppCompatActivity() {
12
13     private val tag = Journaler.tag
14
15     override fun onCreate(savedInstanceState: Bundle?, persistentState: PersistableBundle?) {
16         super.onCreate(savedInstanceState, persistentState)
17         setContentView(R.layout.activity_main)
18         Log.v(tag, "[ ON CREATE ]")
19     }
20
21     override fun onPostCreate(savedInstanceState: Bundle?) {
22         super.onPostCreate(savedInstanceState)
23         Log.v(tag, "[ ON POST CREATE ]")
24     }
25
26     override fun onRestart() {
27         super.onRestart()
28         Log.v(tag, "[ ON RESTART ]")
29     }
30
31     override fun onStart() {
32         super.onStart()
33         Log.v(tag, "[ ON START ]")
34     }
35
36     override fun onResume() {
37         super.onResume()
38         Log.v(tag, "[ ON RESUME ]")
39     }
40
41     override fun onPostResume() {
42         super.onPostResume()
43         Log.v(tag, "[ ON POST RESUME ]")
44     }
45
46     override fun onPause() {
47         super.onPause()
48         Log.v(tag, "[ ON PAUSE ]")
49     }
50
51     override fun onStop() {
52         super.onStop()
53         Log.v(tag, "[ ON STOP ]")
54     }
55
56     override fun onDestroy() {
57         super.onDestroy()
58         Log.v(tag, "[ ON DESTROY ]")
59     }
60
61 }
```

图 2.12

其中，我们在第 18、23、28、33、38 等行上设置了断点。单击 debug 图标，或者选择 Run | Debug app 命令可在调试模式下运行应用程序。相应地，该应用程序将在调试模式下运行。稍等片刻，调试器很快会进入所设置的第一个断点处，如图 2.13 所示。

通过观察可知，第一个进入的方法是 Application 类的 onCreate()方法。接下来检测应用程序是否按照期望结果进入生命周期中的方法。单击 Debugger 面板的 Resume Program 图标，可以看到，针对当前主活动，我们尚未进入 onCreate()方法，而是进入了 Application 类 onCreate()方法之后的 onStart()方法。在此也恭喜读者，你们发现了第一个 Android bug。那么，为什么会出现这种情况？其原因在于，这里采用了错误的 onCreate()方法版本，而非下列代码行：

```
void onCreate(@Nullable Bundle savedInstanceState)
```

并不慎覆写了下列方法：

```
onCreate(Bundle savedInstanceState, PersistableBundle
persistentState)
```

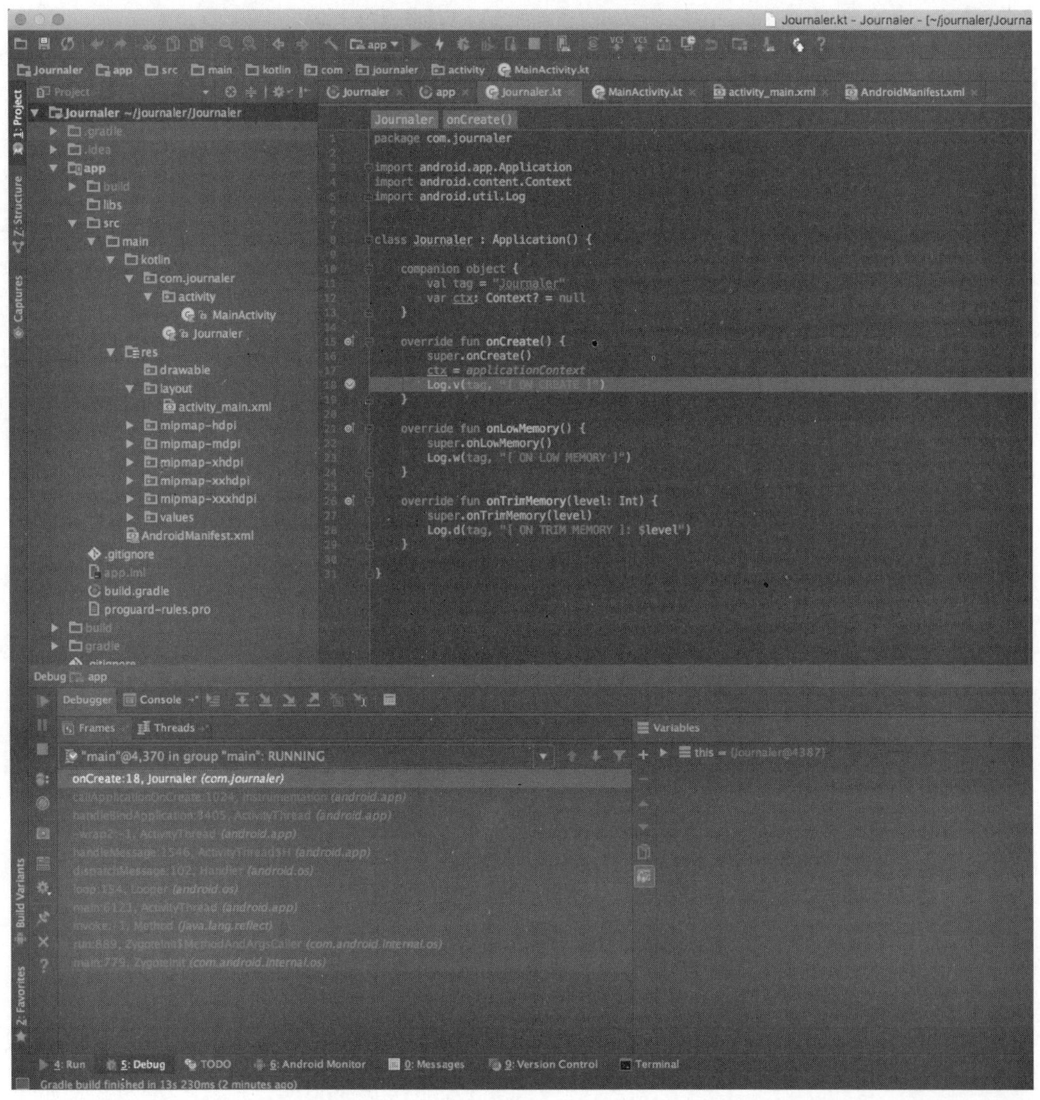

图 2.13

鉴于调试机制，我们发现了问题所在。随后，单击 Debugger 面板中的 Stop 图标可终止调试器并修复代码，如下所示。

```
override fun onCreate(savedInstanceState: Bundle?) {
  super.onCreate(savedInstanceState)
  setContentView(R.layout.activity_main)
  Log.v(tag, "[ ON CREATE 1 ]")
}
override fun onCreate(savedInstanceState: Bundle?,
persistentState: PersistableBundle?) {
  super.onCreate(savedInstanceState, persistentState)
  Log.v(tag, "[ ON CREATE 2 ]")
}
```

在更新了日志消息后，即可跟踪两个 onCreate()方法版本的进入状态。保存更改结果并再次在调试模式下启动应用程序。同时，还应在两个 onCreate()方法处设置断点，并逐一访问断点。当前，我们将以一种期望的方式进入所有的断点。

当查看全部断点时，可单击 View Breakpoints 图标，随后将显示如图 2.14 所示的 Breakpoints 窗口。

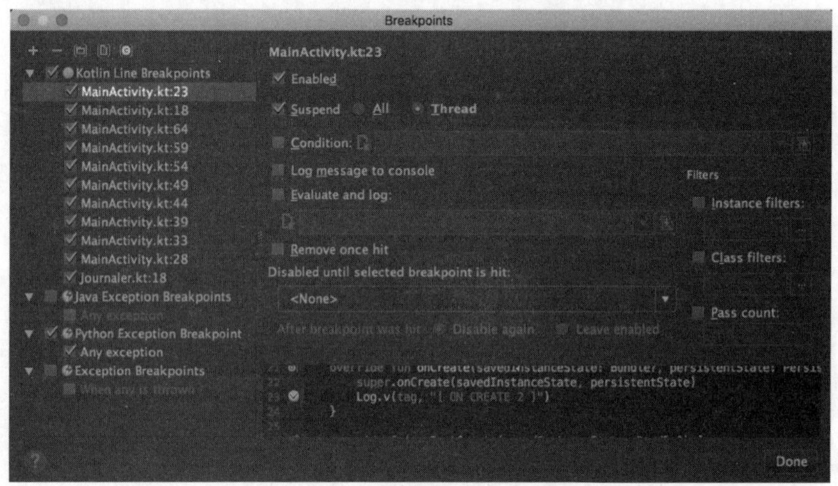

图 2.14

当双击 Breakpoint 时，即可定位在所设置的对应行上。随后，终止调试器。

如果应用程序变得非常庞大，同时还执行了一些开销较大的操作，那么在 Debug 模式下运行该应用程序是非常困难和耗时的。在进入感兴趣的断点之前，我们将损失大量的时间。在调试模式下运行的应用程序其速度通常较慢，对于这一类慢速、庞大的应用程序，如何跳过这一部分操作进而节省宝贵的时间呢？对此，可通过单击 Run 图标或通过选择 Run | Run 'app'命令运行应用程序。此时，应用程序将在部署目标（真实的设备或

模拟器）上启动、运行。通过单击 Attach debugger to Android Process 图标，或者通过选择 Run | Attach debugger to Android 命令，可将调试器绑定至当前应用程序上。随后将显示 Choose Process 窗口，如图 2.15 所示。

双击数据包名称即可选取应用程序进程，随后将显示 Debugger 模板。当前，Debugger 将进入主活动的 main()方法。最后，即可结束 Debugger 操作。

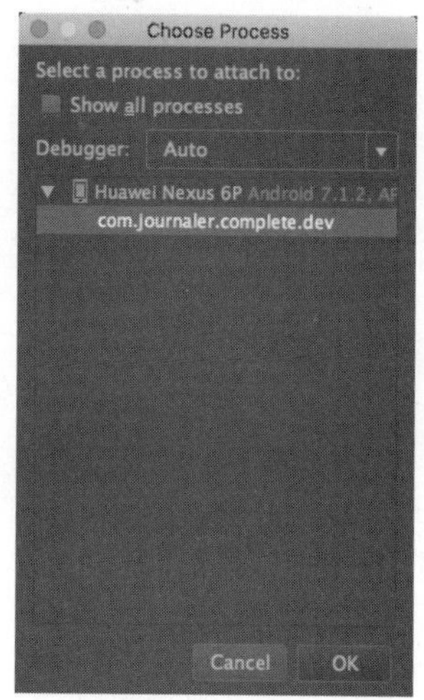

图 2.15

2.5 本 章 小 结

本章介绍了如何在 Android Studio IDE 中，或直接在终端中构建、运行应用程序。此外，本章还对来自模拟器或真实设备的日志消息进行了分析。最后，本章还讨论了调试机制。

第 3 章将集中探讨 UI 组件——屏幕，并讨论如何创建新的屏幕，并向其中加入某些风格化的细节内容。除此之外，第 3 章还将介绍按钮和图像的一些复杂布局。

第 3 章　屏　　幕

通常，配置了普通用户界面的屏幕往往难以令人兴奋。在讨论更为丰富的内容之前，用户需要创建多个屏幕，其中包含开发专业应用程序所必备的所有元素，而这些元素在现代应用程序中普遍存在。第 2 章中曾构建、运行了相关项目，其中所涉及的各种技巧对于本章内容来说依然十分重要。在本章中，我们将向应用程序中加入 UI 元素。

本章主要涉及以下主题：

- 分析模型。
- 定义应用程序活动。
- Android 布局。
- Android Context。
- 片段、片段管理器和栈。
- ViewPager。
- 事务、对话片段和通知。
- 其他重要的 UI 组件。

3.1　分析模型

本节将针对当前应用程序创建所有的屏幕。然而，在构建屏幕之前，首先需要构建和分析模型，以便准确地了解所构建的内容。该模型代表了未加设计的应用程序线框图，仅表示为屏幕布局及其之间的关系。当利用线框图构建模型时，将需要使用相关的线条绘制工具。此处采用了 Pencil，这是一款提供了 GUI 原型的开源应用程序。

图 3.1 显示了当前模型的示意图。

可以看到，图 3.1 中的模型展示了一个相对简单的应用程序，其中包含了一些屏幕。相应地，这些屏幕将包含不同的组件。稍后，我们将在每个屏幕中解释这些组件。

在图 3.1 中，第一个屏幕为 Landing screen，即主应用程序屏幕。每次进入该应用程序时，都将会显示这一屏幕。对此，之前我们已经定义了 MainActivity 类，该项活动即被视为当前屏幕。稍后，还将扩展相应的代码以使该活动与模型所描述的内容吻合。

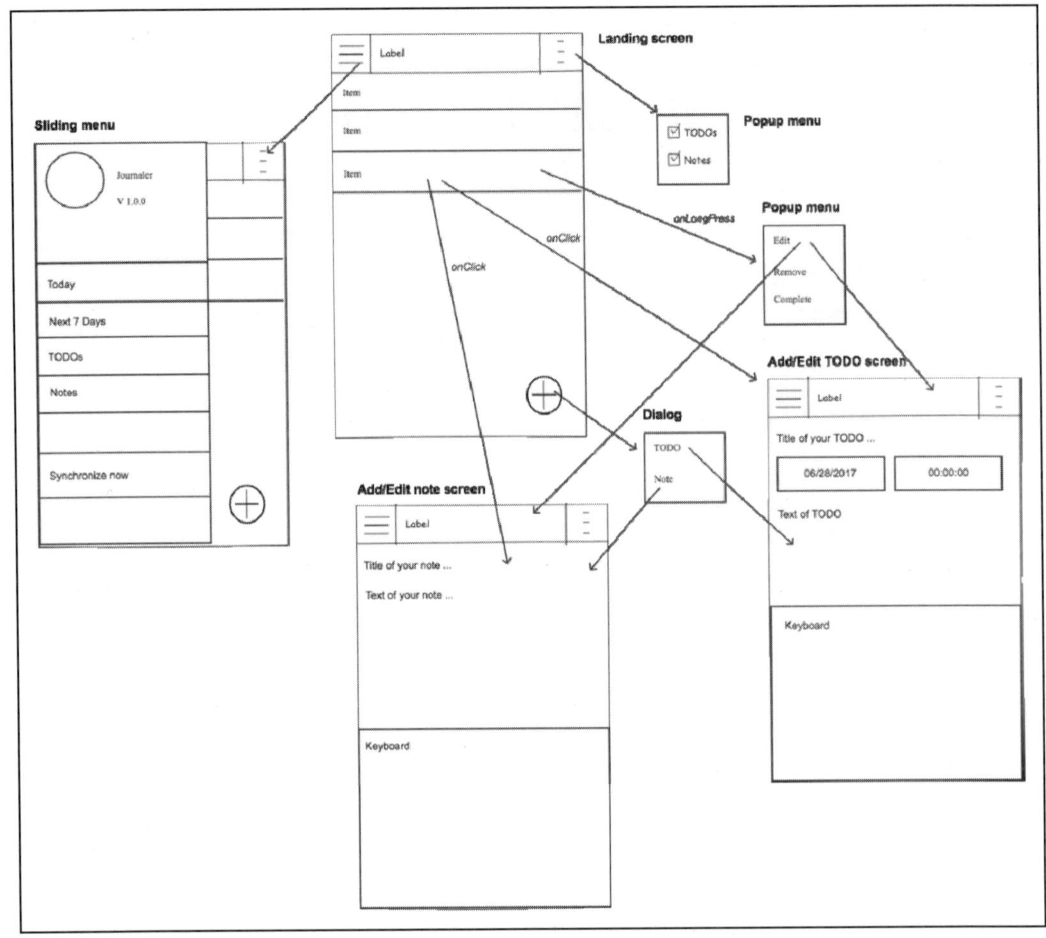

图 3.1

Landing screen 屏幕的中心部分表示为一个列表，其中包含了所创建的所有条目。每个条目将涵盖基本的属性，例如标题、日期和时间。对此，我们将通过相关类型过滤这些条目。具体来说，此处仅可过滤 Note 或 TODO。Note 和 TODO 之间的差别在于，TODO 体现了包含所分配日期和时间的任务。另外，这里还将支持诸如 onLongPress 事件等功能。每个条目上的 onLongPress 事件表示为包含 Edit、Remove 或 Complete 选项的 Popup menu。其中，单击 Edit 将打开屏幕以执行更新操作。

Londing screen 屏幕的右下角设置了一个“+”按钮，该按钮的功能是打开 Dialog 选项，用户可于其中选择是否创建 Note 或 TODO 任务。根据该选项，用户可选择所显示的屏幕之一——Add/Edit note screen 或 Add/Edit TODO screen。

除此之外，Landing screen 屏幕还包含了位于左上角的 Sliding menu 按钮，单击该按钮将打开包含下列条目的 Sliding menu 屏幕：

- ❑ 包含应用程序标题和版本的应用程序图标。
- ❑ Today 按钮，仅过滤分配了当前日期的 TODO 条目。
- ❑ Next 7 Days 按钮，用于过滤分配了 Next 7 Days（包含当前日期）的 TODO 条目。
- ❑ 仅过滤 TODO 条目的 TODO 按钮。
- ❑ 仅过滤 Note 条目的 Note 按钮。

通过单击 Landing screen 屏幕右上角，得到 Popup menu 中的复选框，应用其中的一些过滤器将会对其产生影响。相应地，选中和取消选中这些复选框也将对当前所应用的过滤器加以更改。

Sliding menu 屏幕中的最后一个条目是 Synchronize now，该按钮将引发同步操作，并将所有未同步的条目与后端同步（如果存在）。

接下来将探讨与 Note 和 TODO 创建（或编辑）相关的两个屏幕，具体如下：

- ❑ Add/Edit note screen：该屏幕用于创建新的 Note 或更新现有 Note 的内容。当编辑文本框处于焦点状态时，将开启 Keyboard。考虑到将即刻应用所有更改内容，因而这里未设置保存或更新按钮。当处于该屏幕中时，左上角和右上角的按钮将被禁用。
- ❑ Add/Edit TODO screen：该屏幕用于创建新的 TODO 应用程序，或者更新现有 TODO 的内容。同样，届时将会打开 Keyboard，且未设置保存和更新按钮。另外，左上角和右上角的按钮也处于禁用状态。随后是标题视图，其中的一些按钮用于选取日期和时间。默认状态下，对应内容将被设置为当前日期和时间。相应地，打开 Keyboard 将会凸显按钮。

上述内容介绍了基本的 UI 以及模型的相关功能，接下来将创建一些新的屏幕。

综上所述，下面将定义 3 项活动，具体如下：

- ❑ Landing activity（MainActivty.kt 文件）。
- ❑ Add/Edit note screen。
- ❑ Add/Edit TODO screen。

在 Android 开发过程中，创建一个活动并作为其他活动的父类是一种很常见的做法，进而可减少代码库同时在多个活动中实现共享。在大多数时候，Android 开发人员将其称作 BaseActivity。相应地，我们将定义自己的 BaseActivity 版本。当定义 BaseActivity 新类时，将创建 BaseActivity.kt 文件。这里，应确保新创建的类位于项目的 Activity 包下。

BaseActivity 应扩展 Android SDK 的 FragmentActivity 类，其原因在于，我们将在 MainActivity 类中使用片段（fragment）。此处，片段将与 ViewPager 协同使用，进而在

不同的过滤器（如 Today、Next 7 Days 等）间进行查看。当用户单击 Sliding menu 屏幕中的一项时，ViewPager 将自动滑动至相应的片段位置处（包含了根据所选标准过滤的数据）。相应地，我们将按照下列方式扩展数据包中的 FragmentActivity，即 android.support.v4.app.FragmentActivity。

Android 提供了一种方式进而可支持多个 API 版本。针对于此，这里将使用 support 库中的 FragmentActivity 版本，进而最大化兼容性。当添加 Android support 库时，可在 build.gradle 配置中加入下列指令：

```
compile 'com.android.support:appcompat-v7:26+'
```

由于针对全部活动引入了基类，因而需要对现有的唯一活动实施少量重构。对此，我们将 MainActivity 的 tag 字段移至 BaseActivity 中。鉴于对 BaseActivity 子类的可访问性，因而将更新对 protected 的可见性。

此处希望每个 Activity 类都包含自己的唯一 tag，并通过活动的具体化机制选择其 tag 值。因此，tag 字段变为 abstract，且未分配默认值，如下所示。

```
protected abstract val tag : String
```

除此之外，所有的活动中均涵盖了一些共有的内容，如布局。布局是 Android 通过整型 ID 予以标识的。在 BaseActivity 类中，我们将定义一个 abstract 方法，如下所示。

```
protected abstract fun getLayout(): Int
```

为了进一步优化代码，这里将把 onCreate 从 MainActivity 移到 BaseActivity。我们将传递 getLayout()方法的结果值，而不是直接从 Android 生成的资源中传递布局的 ID。相应地，我们还将移动其他的生命周期覆写方法。

随后可根据上述调整内容更新类，进而构建和运行应用程序，如下所示。

```
BasicActivity.kt:
package com.journaler.activity
import android.os.Bundle
import android.support.v4.app.FragmentActivity
import android.util.Log

abstract class BaseActivity : FragmentActivity() {
  protected abstract val tag : String
  protected abstract fun getLayout(): Int

  override fun onCreate(savedInstanceState: Bundle?) {
    super.onCreate(savedInstanceState)
    setContentView(getLayout())
```

```
  Log.v(tag, "[ ON CREATE ]")
}

override fun onPostCreate(savedInstanceState: Bundle?) {
  super.onPostCreate(savedInstanceState)
  Log.v(tag, "[ ON POST CREATE ]")
}

override fun onRestart() {
  super.onRestart()
  Log.v(tag, "[ ON RESTART ]")
}

override fun onStart() {
  super.onStart()
  Log.v(tag, "[ ON START ]")
}

override fun onResume() {
  super.onResume()
  Log.v(tag, "[ ON RESUME ]")
}

override fun onPostResume() {
  super.onPostResume()
  Log.v(tag, "[ ON POST RESUME ]")
}

override fun onPause() {
  super.onPause()
  Log.v(tag, "[ ON PAUSE ]")
}

override fun onStop() {
  super.onStop()
  Log.v(tag, "[ ON STOP ]")
}

override fun onDestroy() {
  super.onDestroy()
  Log.v(tag, "[ ON DESTROY ]")
}
```

```
}
MainActivity.kt:
package com.journaler.activity
import com.journaler.R

class MainActivity : BaseActivity() {
  override val tag = "Main activity"
  override fun getLayout() = R.layout.activity_main
}
```

接下来将定义其他屏幕。相应地，需要创建一个用于添加、编辑 Note 的屏幕，以及执行相同操作的 TODO 屏幕。需要说明的是，这一类屏幕间涵盖了大量的共有内容，当前唯一的差别在于，TODO 屏幕包含了日期和时间的按钮。相应地，我们将针对屏幕间共享的内容定义一个公共类，每个具体类都将扩展该类。具体来说，此处将定义一个 ItemActivity 类，并确保该类位于 Activity 数据包中。此外，还需定义 NoteActivity 类和 TodoActivity 类。其中，ItemActivity 类扩展 BaseActivity 类，而 NoteActivity 活动类和 TodoActivity 活动类则扩展 ItemActivity 类。其间，将会对某些成员进行覆写。另外，还需要为日志中使用的 tag 提供一些有意义的数值。当分配一个适宜的布局 ID 时，首先需要创建一个布局。

找到我们为主屏幕创建的布局。当前，使用相同的原则创建另外两个布局，具体如下：

- ❑ activity_note.xml 表示为 LinearLayout 类。
- ❑ activity_todo.xml 表示为 LinearLayout 类。

Android 中的任何布局或布局成员都将在构建期间生成的 R 类中以整数形式获得唯一 ID。应用程序的 R 类定义如下：

```
com.journaler.R
```

当访问布局时，可采用下列代码行：

```
R.layout.layout_you_are_interested_in
```

注意，此处采用了静态访问。下面更新具体类并访问布局 ID，如下所示。

```
ItemActivity.kt:
abstract class ItemActivity : BaseActivity()
For now, this class is short and simple.
NoteActivity.kt:
package com.journaler.activity
import com.journaler.R
class NoteActivity : ItemActivity(){
```

```
  override val tag = "Note activity"
  override fun getLayout() = R.layout.activity_note
}
Pay attention on import for R class!

TodoActivity.kt:
package com.journaler.activity
import com.journaler.Rclass TodoActivity : ItemActivity(){
  override val tag = "Todo activity"
  override fun getLayout() = R.layout.activity_todo
}
```

最后一步是注册 view groups 中的屏幕（活动）。对此，打开 manifest 文件并添加下列代码：

```
<activity
  android:name=".activity.NoteActivity"
  android:configChanges="orientation"
  android:screenOrientation="portrait" />

  <activity
    android:name=".activity.TodoActivity"
    android:configChanges="orientation"
    android:screenOrientation="portrait" />
```

其中，NoteActivity 和 TodoActivity 这两个活动均被锁定为 portrait 方向。

至此，我们定义了应用程序屏幕，接下来将利用 UI 组件填充屏幕。

3.2 Android 布局

本节将针对每个屏幕定义相应的布局。Android 中的布局在 XML 文件中加以定义，本节将考查较为常见的布局类型，并利用较常使用的布局组件对其进行填充。

每个布局文件都定义了一个布局类型作为其顶级容器。另外，布局还可包含带有 UI 组件的其他布局，进而实现布局的嵌套。

下列内容显示了较为常用的布局类型：

- ❑ Linear layout：表示为将 UI 组件垂直或水平地以线性顺序对齐。
- ❑ Relative layout：表示为 UI 组件采用彼此相对方式对齐。
- ❑ List view layout：表示为所有条目以列表显示加以组织。

- Grid view layout：表示为所有条目以网格形式加以组织。
- Scroll view layout：表示为内容的高度超过屏幕的实际高度时，可以使用此选项来启用滚动操作。

上述布局元素均为 view groups。每个视图分组还包含了其他视图。view groups 扩展了 ViewGroup 类。从最上方来看，一切均可表示为 View 类。扩展了 View 类（视图）但未扩展 ViewGroup 的类无法包含其他元素（子类），相关示例包括 Button、ImageButton、ImageView 等类。因此，可以定义包含 LinearLayout 的 RelativeLayout，其中包含了垂直或水平对齐的其他多个视图等。

下列内容展示了较为常用的一些视图：

- Button：表示为一个基类，表示一个链接到所定义的 onClick 操作的按钮。
- ImageButton：表示为一个使用图像作为其视觉表达的按钮。
- ImageView：表示为一个视图，并可显示一幅加载自不同来源处的图像。
- TextView：表示为一个视图，其中包含了不可编辑的文本。
- EditText：表示为一个视图，其中包含了可编辑的文本。
- WebView：表示为一个视图，并可显示加载自不同来源处的 HTML 显示页面。
- CheckBox：表示为包含两种状态的选择视图。

每个 View 和 ViewGroup 都支持多种 XML 属性，一些属性仅限于特定的视图类型。此外，全部视图还包含了一些相同的属性，本章稍后将通过屏幕示例着重讨论一些最为常用的视图属性。

当分配一个唯一的标识符并以此通过代码或其他布局成员访问视图时，需要定义一个 ID。当向某个视图分配 ID 时，可采用下列语法：

```
android:id="@+id/my_button"
```

在上述示例中，向视图分配了一个 my_button ID。当从代码中对其进行访问时，可使用下列代码：

```
R.id.my_button
```

其中，R 表示为一个生成后的类，并提供了资源的访问能力。当创建一个按钮的实例时，可采用定义于 Android Activity 类中的 findViewById()方法，如下所示。

```
val x = findViewById(R.id.my_button) as Button
```

鉴于我们选择了 Kotlin，因而可直接对其进行访问，如下所示。

```
my_button.setOnClickListener { ... }
```

注意：

IDE 将询问有关正确导入的问题。请记住，其他布局资源文件也可能包含与定义的名称相同的 ID。在这种情况下，可能会出现导入错误！如果发生这种情况，应用程序将崩溃。

字符串开头的“@”符号表明，XML 解析器应该解析和扩展 ID 字符串的其余部分，并将其标识为 ID 资源。符号“+”表示这是一个新的资源名称。当引用一个 Android 资源 ID 时，不需要使用“+”符号，如下所示。

```
<ImageView
  android:id="@+id/flowers"
  android:layout_width="fill_parent"
  android:layout_height="fill_parent"
  android:layout_above="@id/my_button"
/>
```

下面针对主应用程序屏幕构建 UI。首先设置一些先决条件。在数值资源目录中，创建 dimens.xml 并定义一些将使用的尺寸，如下所示。

```
<?xml version="1.0" encoding="utf-8"?>
<resources>
  <dimen name="button_margin">20dp</dimen>
  <dimen name="header_height">50dp</dimen>
</resources>
```

Android 采用下列单位定义了相应的尺寸维度：

- px（像素）：表示为对应于屏幕上的实际维度。
- in（英寸）：表示为基于屏幕的物理尺寸，也就是说，1 英寸 =2.54 厘米。
- mm（毫米）：表示为基于屏幕的物理尺寸。
- pt（点）：表示为基于屏幕的物理尺寸，即 1 英寸的 1/72。

其他较为重要的单位还包括：

- dp（密度无关像素）：表示为基于屏幕物理密度的抽象单位。相对于 160 DPI 屏幕，一个 dp 是 160 DPI 屏幕上的一个像素。dp 与像素之间的比率会随着屏幕密度的变化而变化，但不一定成正比。
- sp（与比例无关的像素）：sp 类似于 dp 单位，通常用于字体的尺寸。

针对于此，需要创建一个包含于所有屏幕上的标题布局。下面生成一个 activity_header.xml 文件，并按照下列方式进行定义：

```
<?xml version="1.0" encoding="utf-8"?>
```

```
<RelativeLayout xmlns:android=
"http://schemas.android.com/apk/res/android"
android:layout_width="match_parent"
android:layout_height="@dimen/header_height">
<Button
  android:id="@+id/sliding_menu"
  android:layout_width="@dimen/header_height"
  android:layout_height="match_parent"
  android:layout_alignParentStart="true" />

<TextView
  android:layout_centerInParent="true"
  android:id="@+id/activity_title"
  android:layout_width="wrap_content"
  android:layout_height="wrap_content" />

<Button
  android:id="@+id/filter_menu"
  android:layout_width="@dimen/header_height"
  android:layout_height="match_parent"
  android:layout_alignParentEnd="true" />

</RelativeLayout>
```

其中，RelativeLayout 定义为主容器。由于所有元素都相对于父元素彼此定位，所以可使用一些特殊属性来表示这些关系。

对于每个视图，需要设置宽度和高度属性，对应数值如下所示。

❑ 定义于尺寸资源文件中的尺寸，如下所示。

```
android:layout_height="@dimen/header_height"
```

❑ 直接定义的尺寸值，如下所示。

```
android:layout_height="50dp"
```

❑ 匹配父元素的尺寸（match_parent）。

❑ 包装视图的内容（wrap_content）。

接下来，我们将利用子视图填充布局。当前包含 3 个子视图，我们将定义两个按钮和一个文本视图。其中，文本视图将与布局的中心位置对齐；按钮则与布局的边缘对齐——一个按钮与左侧对齐，另一个按钮与右侧对齐。当实现文本视图的中心对齐时，使用了 layout_centerInParent 属性，且传递至其中的数值为布尔值 true。当按钮与布局的左侧对齐时，使用了 layout_alignParentStart 属性；对于右侧，则使用了 layout_alignParentEnd 属性。

每个子元素均包含了所分配的 ID。这一类信息将包含至 MainActivity 中，如下所示。

```
<?xml version="1.0" encoding="utf-8"?>
<LinearLayout xmlns:android=
    "http://schemas.android.com/apk/res/android"
    android:layout_width="match_parent"
    android:layout_height="match_parent"
    android:orientation="vertical">

<include layout="@layout/activity_header" />

<RelativeLayout
    android:layout_width="match_parent"
    android:layout_height="match_parent">
<ListView
    android:id="@+id/items"
    android:layout_width="match_parent"
    android:layout_height="match_parent"
    android:background="@android:color/darker_gray" />

<android.support.design.widget.FloatingActionButton
    android:id="@+id/new_item"
    android:layout_width="wrap_content"
    android:layout_height="wrap_content"
    android:layout_alignParentBottom="true"
    android:layout_alignParentEnd="true"
    android:layout_margin="@dimen/button_margin" />

</RelativeLayout>
</LinearLayout>
```

Main activity 的主容器是 LinearLayout。LinearLayout 的 orientation 则强制设定为如下：

```
android:orientation="vertical"
```

相应地，可分配至其中的值为 vertical 或 horizontal。作为 Main activity 的第一个子元素，此处包含了 activity_header 布局。随后定义了 RelativeLayout，用于填充屏幕的其他内容。

RelativeLayout 包含了两个成员。其中，ListView 将显示所有的条目，并向其中分配了一个背景。当前，我们并未在颜色资源文件中定义自己的颜色，而是采用了 Android 中预定义的颜色。此处最后一个视图则是 FloatingActionButton，该视图也出现于 Gmail Android 应用程序中。按钮将定位于列表上方，屏幕底部的条目则对齐至右侧。除此之外，我们还设置了一个边距，并将按钮从四面包围。

在运行应用程序之前，还需要对其稍作调整。对此，打开 BaseActivity 并按照下列方式更新代码：

```
...
protected abstract fun getActivityTitle(): Int

override fun onCreate(savedInstanceState: Bundle?) {
  super.onCreate(savedInstanceState)
  setContentView(getLayout())
  activity_title.setText(getActivityTitle())
  Log.v(tag, "[ ON CREATE ]")
}
...
```

其中引入了 abstract 方法，并针对每项活动提供了相应的标题。另外，这里还将访问定义于 activity_header.xml 中的 activity_title 视图，该视图包含在当前活动中，并分配相应的方法值。

打开 MainActivity 并覆写下列方法：

```
override fun getActivityTitle() = R.string.app_name
```

随后，可向 ItemActivity 加入相同的代码行。最后，运行应用程序，其主屏幕如图 3.2 所示。

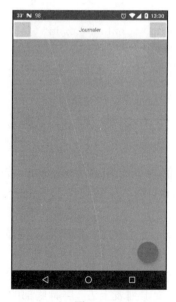

图 3.2

下面针对屏幕的其他内容定义布局。对于 Note、Add/Edit note screen，我们将定义下列布局：

```
<?xml version="1.0" encoding="utf-8"?>
<ScrollView xmlns:android=
  "http://schemas.android.com/apk/res/android"
  android:layout_width="match_parent"
  android:layout_height="match_parent"
  android:fillViewport="true" >

<LinearLayout
  android:layout_width="match_parent"
  android:layout_height="wrap_content"
  android:orientation="vertical">

  <include layout="@layout/activity_header" />

  <EditText
    android:id="@+id/note_title"
    android:layout_width="match_parent"
    android:layout_height="wrap_content"
    android:hint="@string/title"
    android:padding="@dimen/form_padding" />

  <EditText
    android:id="@+id/note_content"
    android:layout_width="match_parent"
    android:layout_height="match_parent"
    android:gravity="top"
    android:hint="@string/your_note_content_goes_here"
    android:padding="@dimen/form_padding" />

</LinearLayout>
</ScrollView>
```

此处引入了 ScrollView 作为布局的上方容器。由于需要填写多行提示信息，因而对应内容可能会低于屏幕的物理限制。对此，可对对应内容执行滚动操作。其间会使用到一个重要的属性，即 fillViewport。该属性将通知容器扩展整个屏幕。同时，全部子元素也将会使用该空间。

3.2.1 使用 EditText 视图

本节引入了 EditText 视图进而可输入可编辑的文本内容，其属性如下：

- hint：该属性定义向用户显示的默认的字符串值。
- padding：该属性定义视图自身与其内容之间的空间。
- gravity：该属性定义内容的方向。在当前示例中，全部文本均位于父视图的上方。

注意：

对于所有的字符串和尺寸，我们在 strings.xml 文件和 dimens.xml 文件中定义了相应的设置项。

字符串资源文件如下所示。

```
<resources>
  <string name="app_name">Journaler</string>
  <string name="title">Title</string>
  <string name="your_note_content_goes_here">Your note content goes
  here.</string>
</resources>
Todos screen will be very similar to this:
<?xml version="1.0" encoding="utf-8"?>
<ScrollView xmlns:android=
  "http://schemas.android.com/apk/res/android"
  android:layout_width="match_parent"
  android:layout_height="match_parent"
  android:fillViewport="true">

<LinearLayout
  android:layout_width="match_parent"
  android:layout_height="wrap_content"
  android:orientation="vertical">

<include layout="@layout/activity_header" />

<EditText
  android:id="@+id/todo_title"
  android:layout_width="match_parent"
  android:layout_height="wrap_content"
  android:hint="@string/title"
  android:padding="@dimen/form_padding" />
```

```
<LinearLayout
  android:layout_width="match_parent"
  android:layout_height="wrap_content"
  android:orientation="horizontal"
  android:weightSum="1">

<Button
  android:id="@+id/pick_date"
  android:text="@string/pick_a_date"
  android:layout_width="0dp"
  android:layout_height="wrap_content"
  android:layout_weight="0.5" />

<Button
  android:id="@+id/pick_time"
  android:text="@string/pick_time"
  android:layout_width="0dp"
  android:layout_height="wrap_content"
  android:layout_weight="0.5" />

</LinearLayout>

<EditText
  android:id="@+id/todo_content"
  android:layout_width="match_parent"
  android:layout_height="match_parent"
  android:gravity="top"
  android:hint="@string/your_note_content_goes_here"
  android:padding="@dimen/form_padding" />
</LinearLayout>
</ScrollView>
```

再次强调，上方容器当前设定为ScrollView。与之前的屏幕比较，当前示例涵盖了一些变化内容，其中加入了容器并加载了日期和时间按钮。另外，对应方向为水平方向。同时，我们还设置了父容器属性weightSum来定义权重值，该值可以被子视图分割，因此每个子视图使用由自身权重定义的空间量。具体来说，weightSum定义为1，第一个按钮的layout_weight为0.5，它将占用50%的水平空间。第二个按钮也设置了相同的值。相应地，视图将被一分为二。在XML文件底部，单击Design进而切换至Design视图。对应的按钮布局如图3.3所示。

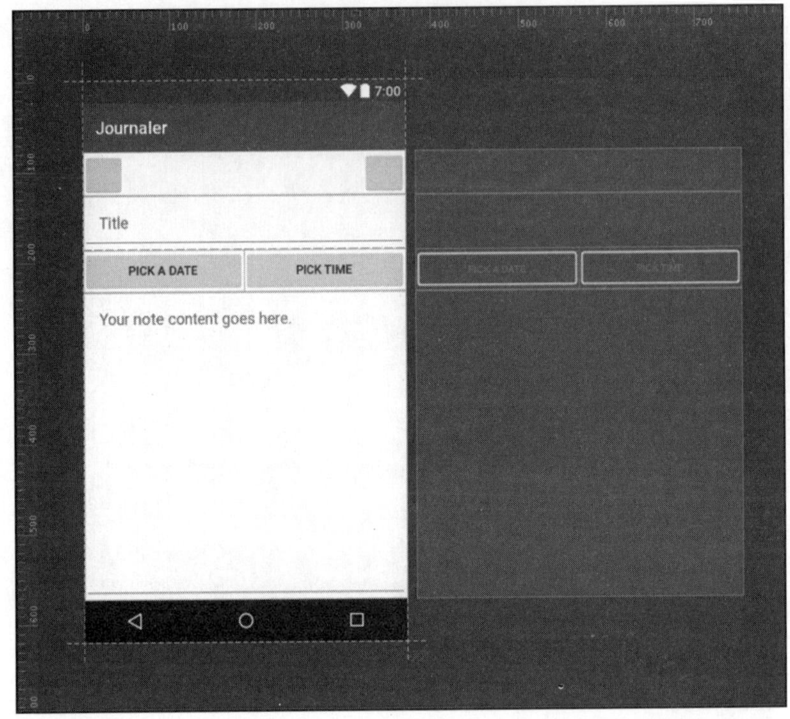

图 3.3

前述内容针对屏幕定义了布局。为了展示屏幕的外观，我们依赖于多种不同的属性，这也仅是可用属性中的一小部分内容。出于完整性考虑，接下来将展示一些在日常开发中常用的其他重要属性。

3.2.2 margin 属性

margin 接收以下列内容为单位的尺寸资源或直接尺寸值：

- layout_margin。
- layout_marginTop。
- layout_marginBottom。
- layout_marginStart。
- layout_marginEnd。

3.2.3 padding 属性

padding 接收以下列内容为单位的尺寸资源或直接尺寸值：

- ❑ padding。
- ❑ paddingTop。
- ❑ paddingBottom。
- ❑ paddingStart。
- ❑ paddingEnd。

3.2.4　检测 gravity 属性

gravity 具体如下：

- ❑ gravity（视图中内容的方向）属性接收以下内容：top、left、right、start、end、center、center_horizontal、center_vertical 等。
- ❑ layout_gravity（父视图中内容的方向）接收以下内容：top、left、right、left、start、end、center、center_horizontal、center_vertical 等。

对于 gravity 属性，还可按照下列方式进行组合：

```
android:gravity="top|center_horizontal"
```

3.2.5　其他属性

前述内容介绍了最为常用的属性。除此之外，其他一些属性也会在特定的场合下发挥自身的作用，其中包括：

- ❑ src：表示为所用的资源，如下所示。

```
android:src="@drawable/icon"
```

- ❑ background：表示为视图的背景，可设置十六进制颜色值或颜色资源数据，如下所示。

```
android:background="#ddff00"
android:background="@color/colorAccent"
```

- ❑ onClick：当用户单击视图（通常是一个按钮）时，将调用该方法。
- ❑ visibility：表示视图的可见性并接收下列参数：gone（不可见且不占用任何布局空间）、invisible（不可见但占用布局空间）和 visible。
- ❑ hint：表示为视图的提示文本，并接收一个字符串值或字符串资源。
- ❑ text：表示为视图的文本，接收一个字符串值或字符串资源。
- ❑ textColor：表示为文本的颜色，可以是十六进制颜色值或颜色资源。

- textSize：表示为所支持单位下的文本的尺寸——单位值或尺寸资源。
- textStyle：表示为定义了相关属性的样式资源，并赋予对应的视图，如下所示。

```
style="@style/my_theme"
...
```

本节讨论了与属性的协调工作方式，若缺乏属性的支持，将无法对 UI 内容进行开发。本章剩余内容将介绍 Android Context。

3.3　理解 Android Context

当前，主屏幕中包含了所定义的布局内容，本节将解释 Android Context，因为刚刚创建的每个屏幕都代表一个 Context 实例。纵观类定义和类扩展，将会发现每个创建的活动都扩展了 Context 类。

Context 代表了应用程序或对象的当前状态，可用于访问应用程序的特定类或资源。例如，考查下列代码行：

```
resources.getDimension(R.dimen.header_height)
getString(R.string.app_name)
```

这里所展示的访问操作是由 Context 类提供的，并显示了所扩展的活动。当发布一项活动、启动一个服务或发送广播消息时，需要使用 Context。在适当的时候，我们还将具体展示此类方法的应用方式。此前已经谈到，Android 应用程序的每个屏幕（Activity）代表了一个 Context 实例。另外，活动不仅仅是上下文类，除了活动之外，还包含服务上下文类型。

Android Context 包含以下功能：

- 显示一个对话。
- 启动一项活动。
- 扩展布局。
- 启动一项服务。
- 绑定服务。
- 发送广播消息。
- 注册广播消息。
- 加载资源（前述示例已对此有所展示）。

Context 是 Android 中的一项较为重要的内容，也是较常使用的框架类之一。另外，本章稍后还将讲述其他 Context 类。在详细介绍之前，下面首先考查片段（fragment）。

3.4　理解片段

如前所述，主屏幕的中央部分将包含过滤后的条目列表。此处希望某些页面应用一组不同的过滤器。用户可以向左或向右滑动来更改过滤后的内容，并浏览以下页面：

- 所显示的全部内容。
- 基于 Today 的条目。
- 基于 Next 7 Days 的条目。
- 仅 Note。
- 仅 TODO。

当实现上述功能时，需要定义片段。那么，片段的含义和功能又是什么？

片段是 Activity 实例接口的一部分，我们可使用片段创建多个屏幕，以及包含视图分页机制的多个屏幕。

与活动类似，片段包含了自身的生命周期。图 3.4 显示了相应的片段生命周期。

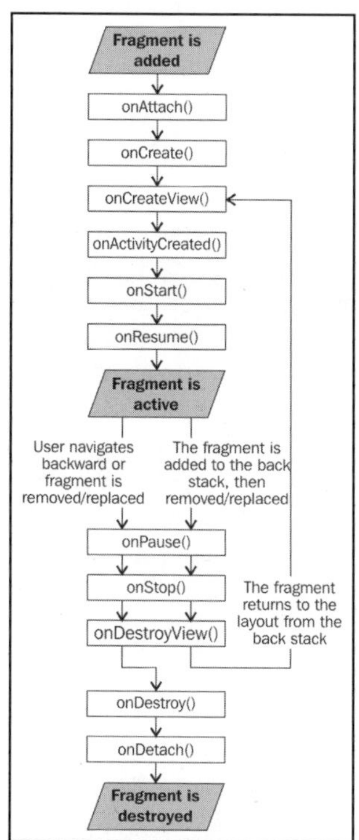

图 3.4

此外，还存在一些活动未曾涉及的一些附加方法，具体如下：

- onAttach()：当片段与某项活动关联时执行该方法。
- onCreateView()：该方法实例化并返回片段的视图实例。
- onActivityCreated()：当某项活动的 onCreate()被执行时，将执行该方法。
- onDestroyView()：当视图被销毁时，该方法将被执行。这对于某些清空操作来说十分方便。
- onDetach()：当片段与某项活动无关联时将执行该方法。为了进一步展示片段的应用，此处将 MainActivity 的中心部分置入某个独立片段中，稍后将其移至 ViewPager 中，并向其中添加更多的页面。

下面创建一个名为 fragment 的新数据包，并于随后定义一个名为 BaseFragment 的新类。根据当前示例更新 BaseFragment 类，如下所示。

```
package com.journaler.fragment

import android.os.Bundle
import android.support.v4.app.Fragment
import android.util.Log
import android.view.LayoutInflater
import android.view.View
import android.view.ViewGroup

abstract class BaseFragment : Fragment() {
  protected abstract val logTag : String
  protected abstract fun getLayout(): Int

  override fun onCreateView(
    inflater: LayoutInflater?, container: ViewGroup?,
    savedInstanceState: Bundle?
    ): View? {
      Log.d(logTag, "[ ON CREATE VIEW ]")
      return inflater?.inflate(getLayout(), container, false)
  }

  override fun onPause() {
    super.onPause()
    Log.v(logTag, "[ ON PAUSE ]")
  }

  override fun onResume() {
    super.onResume()
```

```
    Log.v(logTag, "[ ON RESUME ]")
  }

  override fun onDestroy() {
    super.onDestroy()
    Log.d(logTag, "[ ON DESTROY ]")
  }

}
```

此处需要注意以下导入语句：

```
import android.support.v4.app.Fragment
```

由于需要得到最大的兼容性，因而将从 Android 支持库中导入片段。

可以看出，这里所做的工作类似于处理 BaseActivity。接下来创建一个新的片段，即名为 ItemsFragment 的类，并根据当前示例更新其代码：

```
package com.journaler.fragment
import com.journaler.R

class ItemsFragment : BaseFragment() {
  override val logTag = "Items fragment"
  override fun getLayout(): Int {
    return R.layout.fragment_items
  }
}
```

这里引入了一个新的布局，它实际上包含了 activity_main 中的列表视图。下面创建一个名为 fragment_items 的新布局资源：

```
<?xml version="1.0" encoding="utf-8"?>
<RelativeLayout xmlns:android=
  "http://schemas.android.com/apk/res/android"
  android:layout_width="match_parent"
  android:layout_height="match_parent">

<ListView
  android:id="@+id/items"
  android:layout_width="match_parent"
  android:layout_height="match_parent"
  android:background="@android:color/darker_gray" />

<android.support.design.widget.FloatingActionButton
```

```
  android:id="@+id/new_item"
  android:layout_width="wrap_content"
  android:layout_height="wrap_content"
  android:layout_alignParentBottom="true"
  android:layout_alignParentEnd="true"
  android:layout_margin="@dimen/button_margin" />

</RelativeLayout>
```

不难发现，这部分内容取自 activity_main 布局。取而代之的是，可将下列代码置于 activity_main 布局中：

```
<?xml version="1.0" encoding="utf-8"?>
<LinearLayout xmlns:android=
  "http://schemas.android.com/apk/res/android"
  android:layout_width="match_parent"
  android:layout_height="match_parent"
  android:orientation="vertical">
<include layout="@layout/activity_header" />

<FrameLayout
  android:id="@+id/fragment_container"
  android:layout_width="match_parent"
  android:layout_height="match_parent" />
</LinearLayout>
```

其中，FrameLayout 将作为 fragment 容器。为了进一步展示 fragment_containerFrameLayout 中的新片段，按照以下方式更新 MainActivity 的代码：

```
class MainActivity : BaseActivity() {

  override val tag = "Main activity"
  override fun getLayout() = R.layout.activity_main
  override fun getActivityTitle() = R.string.app_name

  override fun onCreate(savedInstanceState: Bundle?) {
    super.onCreate(savedInstanceState)
    val fragment = ItemsFragment()
    supportFragmentManager
            .beginTransaction()
            .add(R.id.fragment_container, fragment)
            .commit()
  }
}
```

这里，我们访问了 supportFragmentManager。如果未选择使用 Android 支持库，则需要使用 fragmentManager。随后将启动片段事务，并向其中添加一个新的片段实例，该实例将与 fragment_container FrameLayout 关联。相应地，commit()方法负责执行该事务。如果现在运行这一应用程序，将无法体验到任何变化内容；但如果查看日志内容，则会看到片段的生命周期，如下所示。

```
V/Journaler: [ ON CREATE ]
V/Main activity: [ ON CREATE ]
D/Items fragment: [ ON CREATE VIEW ]
V/Main activity: [ ON START ]
V/Main activity: [ ON POST CREATE ]
V/Main activity: [ ON RESUME ]
V/Items fragment: [ ON RESUME ]
V/Main activity: [ ON POST RESUME ]
```

这里向接口中添加了简单的片段。在 3.4.1 节中，读者还将学习与片段管理器及其功能相关的更多内容，并于随后创建一个 ViewPager.

3.4.1　片段管理器

负责与当前活动中的片段交互的组件称作片段管理器。通过以下两个不同的导入语句，即可使用 FragmentManager。

- android.app.FragmentManager。
- android.support.v4.app.Fragment。

这里推荐从 Android 支持库中进行导入。

当执行一系列的编辑操作时，可利用 beginTransaction()方法启动片段事务，这将返回一个事务实例。在添加片段时（通常是首先执行添加操作），可采用 add()方法。该方法接收相同的参数，但会替换当前的参数（如果已经添加）。如果计划向后浏览片段，则需要使用 addToBackStack()方法将事务添加到回退栈。该方法接收一个 name 参数或者 null（如果不希望赋予名称）。

最后，可通过执行 commit()方法来调度事务，并调度应用程序主线程上的操作。当主线程处于就绪状态时，当前事务将被执行。当规划和实现代码时，需要对此予以考虑。

3.4.2　片段栈

当展示片段和回退栈示例时，还需要进一步扩展应用程序。这里将创建一个片段来显示包含文本 Lorem ipsum 的用户手册。对此，首先需要创建一个新的片段，随后创建一个名为 fragment_manual 的新布局，并通过下列方式更新当前示例中的布局：

```
<?xml version="1.0" encoding="utf-8"?>
<LinearLayout xmlns:android=
  "http://schemas.android.com/apk/res/android"
  android:layout_width="match_parent"
  android:layout_height="match_parent"
  android:orientation="vertical">

<TextView
  android:layout_width="match_parent"
  android:layout_height="match_parent"
  android:layout_margin="10dp"
  android:text="@string/lorem_ipsum_sit_dolore"
  android:textSize="14sp" />
</LinearLayout>
```

这是一个简单的布局，包含横跨整个父视图的文本视图。使用该布局的片段称作 ManualFragment。下面针对该片段定义一个类，并确保涵盖以下内容：

```
package com.journaler.fragment
import com.journaler.R

class ManualFragment : BaseFragment() {
  override val logTag = "Manual Fragment"
  override fun getLayout() = R.layout.fragment_manual
}
```

最后，将其添加至片段回退栈中，并按照下列方式更新 MainActivity 的 onCreate()方法：

```
override fun onCreate(savedInstanceState: Bundle?) {
  super.onCreate(savedInstanceState)
  val fragment = ItemsFragment()
  supportFragmentManager
            .beginTransaction()
            .add(R.id.fragment_container, fragment)
            .commit()
  filter_menu.setText("H")
  filter_menu.setOnClickListener {
    val userManualFrg = ManualFragment()
    supportFragmentManager
              .beginTransaction()
              .replace(R.id.fragment_container, userManualFrg)
              .addToBackStack("User manual")
```

```
                .commit()
  }
}
```

构建并运行当前应用程序。随后，右上方标题按钮将包含一个标签 H，单击该标签，包含 Lorem ipsum 文本的片段将填充对应的视图；单击回退按钮，则片段消失。这意味着，我们已经成功地从回退栈中添加、移除了片段。

接下来将要尝试另一项操作，即连续两、三次单击同一个按钮。具体来说，反复单击回退按钮后，将遍历回退栈，直至到达第一个片段。随后，若再次单击回退按钮，将离开当前应用程序。对此，可查看日志内容以获取相关信息。

读者是否还记得执行生命周期方法的顺序？每次新片段添加至顶部时，下方的片段均会处于暂停状态。

3.5　创建视图分页器

如前所述，数据条目应可显示于可滑动的多个页面上。针对于此，需要使用 ViewPager。ViewPager 支持在不同的片段间进行滑动（成为片段集合的一部分内容）。对此，可对之前的代码进行适当的调整。打开 activity_main 布局，并按照下列方式对其进行更新：

```
<?xml version="1.0" encoding="utf-8"?>
<LinearLayout xmlns:android=
  "http://schemas.android.com/apk/res/android"
  android:layout_width="match_parent"
  android:layout_height="match_parent"
  android:orientation="vertical">
<android.support.v4.view.ViewPager xmlns:android=
  "http://schemas.android.com/apk/res/android"
  android:id="@+id/pager"
  android:layout_width="match_parent"
  android:layout_height="match_parent" />

</LinearLayout>
```

这里并不打算采用 FrameLayout，而是置入 ViewPager 视图。随后，打开 MainActivity 类并按照下列方式对其进行更新：

```
class MainActivity : BaseActivity() {
  override val tag = "Main activity"
```

```
  override fun getLayout() = R.layout.activity_main
  override fun getActivityTitle() = R.string.app_name

  override fun onCreate(savedInstanceState: Bundle?) {
    super.onCreate(savedInstanceState)
    pager.adapter = ViewPagerAdapter(supportFragmentManager)
  }

  private class ViewPagerAdapter(manager: FragmentManager) :
  FragmentStatePagerAdapter(manager) {
    override fun getItem(position: Int): Fragment {
      return ItemsFragment()
    }

    override fun getCount(): Int {
      return 5
    }
  }
}
```

当前主要工作是针对分页器定义 adapter，同时需要扩展 FragmentStatePagerAdapter 类。相应地，其构造方法接收用于处理片段事务的片段管理器。其间，还需要重载返回片段实例的 getItem()方法，以及返回全部期望片段数量的 getCount()方法。代码其余部分则较为清晰——访问页面器（利用所分配的 ViewPager 的 ID），并向其分配新的适配器实例。

运行当前应用程序，并尝试左、右方向滑动。在滑动过程中，则观察 Logcat 和生命周期日志的输出结果。

3.6　利用渐变效果实现动画

当模拟片段间的动画效果时，需要向事务实例分配某些动画资源。回忆一下，在开始片段事务之前，需要获得一个事务实例。随后，可访问该实例，并执行下列方法：

```
setCustomAnimations (int enter, int exit, int popEnter, int popExit)
```

或

```
setCustomAnimations (int enter, int exit)
```

其中，每个参数均代表该事务中所用的动画元素。当然，也可定义自己的动画资源，或者使用预定义资源，如图 3.5 所示。

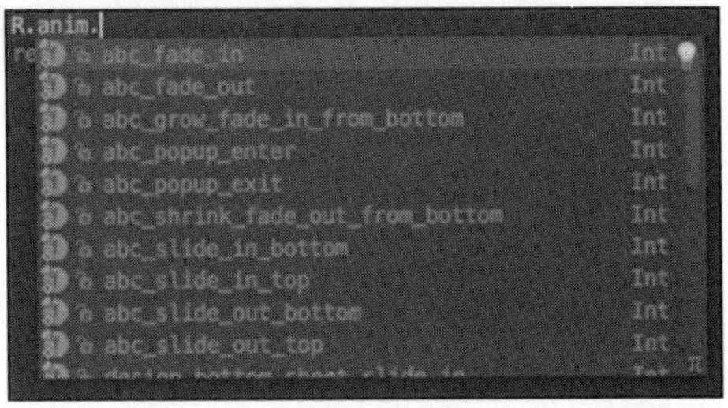

图 3.5

3.7　对话框片段

如果希望片段浮动于应用程序 UI 的上方，则可以使用 DialogFragment。全部工作即是定义该片段，这与之前的做法较为类似，即定义扩展 DialogFragment 的类，并重载定义对应布局的 onCreateView()方法。另外，还需要重载 onCreate()方法。最后，可通过下列方式对其予以显示：

```
val dialog = MyDialogFragment()
dialog.show(supportFragmentManager, "dialog")
```

在该示例中，我们向片段管理器中传递了事务的实例及其名称。

3.8　通　　知

如果向终端用户显示的内容相对短小，则可尝试采用通知，而非对话框。对此，可通过多种方式定制通知，本节将讨论较为基本的定制方法。通知的构建和显示过程较为简单，稍后将对此加以讨论，对应类还会在后续章节中多次遇到。

接下来首先展示如何使用通知，具体步骤如下：

（1）定义一个 notificationBuilder，并传递一个小图标、内容标题和内容文本，如下所示。

```
val notificationBuilder = NotificationCompat.Builder(context)
        .setSmallIcon(R.drawable.icon)
        .setContentTitle("Hello!")
        .setContentText("We love Android!")
```

（2）针对应用程序活动定义 Intent（第 4 章还将对此予以解释），如下所示。

```
val result = Intent(context, MyActivity::class.java)
```

（3）定义栈构造器对象，该对象包含了针对对应活动的回退栈，如下所示。

```
val builder = TaskStackBuilder.create(context)
```

（4）针对当前意图（intent）添加回退栈，如下所示。

```
builder.addParentStack(MyActivity::class.java)
```

（5）在栈顶添加当前意图，如下所示。

```
builder.addNextIntent(result)
val resultPendingIntent = builder.getPendingIntent(
  0,PendingIntent.FLAG_UPDATE_CURRENT )Define ID for the
  notification and notify:
val id = 0
notificationBuilder.setContentIntent(resultPendingIntent)
val manager = getSystemService(NOTIFICATION_SERVICE) as
NotificationManager
manager.notify(id, notificationBuilder.build())
```

3.9　其他重要组件

Android Framework 内容丰富且功能强大。之前曾介绍了 View 类，这也是最为常用的类之一。但是，还存在许多 View 类本章并未涉及，后续章节还会进一步加以讨论。下面对此稍作介绍：

- ConstraintLayout：可通过灵活的方式查看和定位子元素。
- CoordinatorLayout：FrameLayout 的高级版本。
- SurfaceView：该视图用于绘制操作（尤其是对性能要求较高时）。
- VideoView：用于设置、播放视频内容。

3.10　本 章 小 结

本章学习了如何创建可被划分为多个部分的屏幕，当前，读者可创建包含按钮和图像的复杂布局。除此之外，本章还介绍了如何创建对话框和通知。第 4 章将连接所有的屏幕和浏览操作。

第 4 章　连接屏幕流

目前，我们已经步入应用程序开发的一个重要阶段——连接屏幕。

第 3 章曾创建了屏幕，本章将利用 Android 的强大框架对其进行连接，以进一步丰富当前工作和 UI 内容。

本章主要涉及以下主题：

- ❑ 创建应用程序工具栏。
- ❑ 使用导航抽屉。
- ❑ 理解 Android 意图。
- ❑ 在活动和片段之间传递信息。

4.1　创建应用程序工具栏

本章将继续讨论 Android 应用程序开发工作。截至目前，我们构建了应用程序的基础内容、定义了 UI 的基本形式，并创建了主屏幕。然而，此类屏幕并未处于连接状态。本章将尝试对屏幕进行连接，进而生成奇妙的交互效果。

考虑到所有工作均始于 MainActivity 类，对此，本节将实施一些改进工作，并于随后设置相关操作以触发其他屏幕。相应地，我们需要通过应用程序工具栏对其进行封装。这里，工具栏是 UI 中的部分内容，常用于访问应用程序的其他部分，并通过交互元素提供视觉结构。之前曾创建了一个工具栏，但它并不是一般意义上的工具栏。具体来说，它仅是一个经调整的工具栏，而非标准的 Android 应用程序工具栏。

本节将展示如何创建真正意义上的应用程序工具栏。

首先，需要将顶级活动替换为 AppCompatActivity；另外还需要访问应用程序工具栏所需的各种特性。AppCompatActivity 将把这一类附加特性添加至标准的 FragmentActivity 中。BaseActivity 的定义方式如下所示。

```
abstract class BaseActivity : AppCompatActivity() {
...
```

接下来将更新所用的主题应用程序，进而可使用应用程序工具栏。对此，打开 Android Manifest，并按照如下方式设置新的特性：

```
...
<application
  android:name=".Journaler"
  android:allowBackup="false"
  android:icon="@mipmap/ic_launcher"
  android:label="@string/app_name"
  android:roundIcon="@mipmap/ic_launcher_round"
  android:supportsRtl="true"
  android:theme="@style/Theme.AppCompat.Light.NoActionBar">
...
```

下面打开 activity_main 布局，移除头部包含的指令并添加 Toolbar，如下所示。

```
<?xml version="1.0" encoding="utf-8"?>
<LinearLayout xmlns:android=
  "http://schemas.android.com/apk/res/android"
  android:layout_width="match_parent"
  android:layout_height="match_parent"
  android:orientation="vertical">

<android.support.v7.widget.Toolbar
  android:id="@+id/toolbar"
  android:layout_width="match_parent"
  android:layout_height="50dp"
  android:background="@color/colorPrimary"
  android:elevation="4dp" />

<android.support.v4.view.ViewPager
  android:id="@+id/pager"
  android:layout_width="match_parent"
  android:layout_height="match_parent" />

</LinearLayout>
```

针对所有的布局进行相同的调整，在操作完毕后，更新 BaseActivity 代码以使用新的 Toolbar。此时，onCreate()方法如下所示。

```
override fun onCreate(savedInstanceState: Bundle?) {
  super.onCreate(savedInstanceState)
  setContentView(getLayout())
  setSupportActionBar(toolbar)
  Log.v(tag, "[ ON CREATE ]")
}
```

此处丢失了标题头中的按钮，别担心，我们会把它们找回来的！下面将创建一个菜单（而非按钮），用于处理相关操作。在 Android 中，菜单表示为一个界面，并可管理相关条目；另外，用户也可定义自己的菜单资源。在/res 目录中，生成一个名为 menu 的文件夹，右击 menu 文件夹，选择 New | New menu resource 文件，并将其命名为 main。随后打开新的 XML 文件，并通过下列方式更新其内容：

```
<?xml version="1.0" encoding="utf-8"?>
<menu xmlns:android="http://schemas.android.com/apk/res/android"
xmlns:app="http://schemas.android.com/apk/res-auto">

<item
  app:showAsAction="ifRoom"
  android:orderInCategory="1"
  android:id="@+id/drawing_menu"
  android:icon="@android:drawable/ic_dialog_dialer"
  android:title="@string/mnu" />

<item
  app:showAsAction="ifRoom"
  android:orderInCategory="2"
  android:id="@+id/options_menu"
  android:icon="@android:drawable/arrow_down_float"
  android:title="@string/mnu" />
</menu>
```

这里设置了公共属性、图标和对应的顺序。若确保图标处于可见状态，可采用如下方式：

```
app:showAsAction="ifRoom"
```

据此，如果存在合适的空间，菜单中的条目将可下拉显示；否则，它们将通过快捷菜单予以访问。另外，Android 中其他可供选择的空间选项还包括：

- Always：该按钮将一直置于应用程序工具栏中。
- Never：该按钮将不会置于应用程序工具栏中。
- collapseAction View：该按钮将作为一个微件（widget）予以显示。
- withText：该按钮可显示文本内容。

当向应用程序工具栏中分配菜单时，可向 BaseActivity 添加以下内容：

```
override fun onCreateOptionsMenu(menu: Menu): Boolean {
 menuInflater.inflate(R.menu.main, menu)
 return true
}
```

最后，将相关操作连接至菜单项，并通过添加下列代码来扩展 MainActivity：

```
override fun onOptionsItemSelected(item: MenuItem): Boolean {
  when (item.itemId) {
    R.id.drawing_menu -> {
      Log.v(tag, "Main menu.")
      return true
    }
    R.id.options_menu -> {
      Log.v(tag, "Options menu.")
      return true
    }
    else -> return super.onOptionsItemSelected(item)
  }
}
```

这里重载了 onOptionsItemSelected()方法，同时处理了菜单项 ID 的各种情况。在每次选择过程中，均添加了一条日志消息。此时运行上述应用程序，则图 4.1 显示了相应的菜单项。

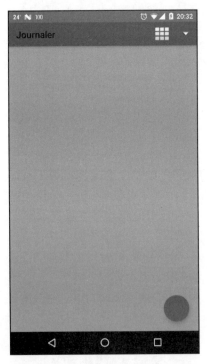

图 4.1

在每个菜单项上单击多次并观察 Logcat 的输出结果，相关日志内容如下所示。

```
V/Main activity: Main menu.
V/Main activity: Options menu.
V/Main activity: Options menu.
V/Main activity: Options menu.

V/Main activity: Main menu.

V/Main activity: Main menu.
```

至此，我们成功地将标题切换到应用程序工具栏，它与应用程序线框图中的标题有很大的不同。当前，这一问题并不重要，后续章节还将实现一些较为重要的样式设计；相应地，应用程序工具栏将随之产生变化。

4.2 节将介绍导航抽屉，并尝试整合应用程序的导航功能。

4.2　使用导航抽屉

如果读者还记得的话，模型中提供了过滤数据（Note 和 Todo）的链接，我们将通过导航抽屉对此进行过滤。当前，应用程序中一般均采用了导航抽屉。导航抽屉是一项 UI 功能，并可显示应用程序的导航选项。当定义导航抽屉时，需要在布局中设置 DrawerLayout 视图。对此，打开 activity_main 并按照下列方式进行修改：

```
<?xml version="1.0" encoding="utf-8"?>
<android.support.v4.widget.DrawerLayout xmlns:android=
  "http://schemas.android.com/apk/res/android"
  android:id="@+id/drawer_layout"
  android:layout_width="match_parent"
  android:layout_height="match_parent">

<LinearLayout
  android:layout_width="match_parent"
  android:layout_height="match_parent"
  android:orientation="vertical">

<android.support.v7.widget.Toolbar
  android:id="@+id/toolbar"
  android:layout_width="match_parent"
  android:layout_height="50dp"
  android:background="@color/colorPrimary"
```

```
  android:elevation="4dp" />

<android.support.v4.view.ViewPager xmlns:android=
  "http://schemas.android.com/apk/res/android"
  android:id="@+id/pager"
  android:layout_width="match_parent"
  android:layout_height="match_parent" />

</LinearLayout>

<ListView
    android:id="@+id/left_drawer"
    android:layout_width="240dp"

    android:layout_height="match_parent"
    android:layout_gravity="start"
    android:background="@android:color/darker_gray"
    android:choiceMode="singleChoice"
    android:divider="@android:color/transparent"
    android:dividerHeight="1dp" />
</android.support.v4.widget.DrawerLayout>
```

屏幕的主要内容应为 DrawerLayout 的第一个子元素。导航抽屉采用了第二个子元素作为抽屉内容，在当前示例中为 ListView。当通知导航抽屉导航项处于左侧或右侧时，可使用 layout_gravity 属性。如果打算使用定位于右侧的导航抽屉，可将该属性值设置为 end。

当前，我们拥有一个空的导航抽屉，且需要通过某些按钮对其进行填充。针对每个导航项，下面创建一个布局文件，将其命名为 adapter_navigation_drawer，并于其中仅通过 1 个按钮实现简单的线性布局，如下所示。

```
<?xml version="1.0" encoding="utf-8"?>
<LinearLayout xmlns:android=
  "http://schemas.android.com/apk/res/android"
  android:layout_width="match_parent"
  android:layout_height="match_parent"
  android:orientation="vertical">

<Button
  android:id="@+id/drawer_item"
  android:layout_width="match_parent"
  android:layout_height="wrap_content" />

</LinearLayout>
```

接下来创建一个名为 navigation 的新数据包，并于其中定义一个新的 Kotlin data 类，如下所示。

```
package com.journaler.navigation
data class NavigationDrawerItem(
  val title: String,
  val onClick: Runnable
)
```

这里定义了一个抽屉项实体，下面定义另一个类，如下所示。

```
class NavigationDrawerAdapter(
  val ctx: Context,
  val items: List<NavigationDrawerItem>
) : BaseAdapter() {

  override fun getView(position: Int, v: View?, group: ViewGroup?):
  View {
    val inflater = LayoutInflater.from(ctx)
    var view = v
    if (view == null) {
      view = inflater.inflate(
        R.layout.adapter_navigation_drawer, null
      ) as LinearLayout
    }

    val item = items[position]
    val title = view.findViewById<Button>(R.id.drawer_item)
    title.text = item.title
    title.setOnClickListener {
      item.onClick.run()
    }

    return view
  }

  override fun getItem(position: Int): Any {
    return items[position]
  }

  override fun getItemId(position: Int): Long {
    return 0L
  }
```

```
  override fun getCount(): Int {
    return items.size
  }

}
```

上述类扩展了 Android 的 BaseAdapter，并重载了适配器所需的相关方法，进而提供了视图实例。适配器创建的全部视图将被分配，以扩展导航抽屉中的 ListView。

最后，我们将分配该适配器。对此，需要执行下列代码以更新 MainActivity 类：

```
class MainActivity : BaseActivity() {
  ...
  override fun onCreate(savedInstanceState: Bundle?) {
    super.onCreate(savedInstanceState)
    pager.adapter = ViewPagerAdapter(supportFragmentManager)

    val menuItems = mutableListOf<NavigationDrawerItem>()
    val today = NavigationDrawerItem(
      getString(R.string.today),
      Runnable {
        pager.setCurrentItem(0, true)
      }
    )

    val next7Days = NavigationDrawerItem(
      getString(R.string.next_seven_days),
        Runnable {
          pager.setCurrentItem(1, true)
        }
    )

    val todos = NavigationDrawerItem(
      getString(R.string.todos),
        Runnable {
            pager.setCurrentItem(2, true)
        }
    )

    val notes = NavigationDrawerItem(
      getString(R.string.notes),
        Runnable {
```

```
            pager.setCurrentItem(3, true)
        }
    )

    menuItems.add(today)
    menuItems.add(next7Days)
    menuItems.add(todos)
    menuItems.add(notes)

    val navgationDraweAdapter =
    NavigationDrawerAdapter(this, menuItems)
    left_drawer.adapter = navgationDraweAdapter
  }

  override fun onOptionsItemSelected(item: MenuItem): Boolean {
    when (item.itemId) {
      R.id.drawing_menu -> {
        drawer_layout.openDrawer(GravityCompat.START)
        return true
      }
      R.id.options_menu -> {
        Log.v(tag, "Options menu.")
        return true
      }
      else -> return super.onOptionsItemSelected(item)
    }
  }
}
```

上述代码示例实例化了多个 NavigationDrawerItem 实例，随后向将要执行的按钮和 Runnable 操作分配了一个标题。其中，每个 Runnable 标题将跳至视图分页器的特定页面。我们将所有实例作为一个可变列表传递给适配器。另外，还需要注意的是，此处还修改了 drawing_menu 项对应的代码行，单击该项后将展开导航抽屉。接下来，可执行以下各项操作步骤：

（1）构建、运行应用程序。

（2）单击主屏幕右上方位置处的菜单按钮，或者通过左、右滑动展开导航抽屉。

（3）单击按钮。

（4）可以看到，视图分页器在导航抽屉下方实现了页面位置的动画效果。

对应结果如图 4.2 所示。

图 4.2

4.3 连接活动

如前所述，除了 MainActivity 之外，还存在其他活动。在当前应用程序中，可生成相应的活动以创建/编辑 Note 和 TODO。此处的目标是，将其连接至按钮单击事件上；随后，当用户单击该按钮时，将会打开相应的屏幕。对此，首先定义一个 enum，表示活动中执行的操作。当开启活动时，即可查看、创建、更新 Note 或 Todo。下面创建一个名为 model 的新数据包，以及包含 MODE 名称的 enum，并确保涵盖下列 enum 值：

```
enum class MODE(val mode: Int) {
  CREATE(0),
  EDIT(1),
  VIEW(2);

  companion object {
    val EXTRAS_KEY = "MODE"

    fun getByValue(value: Int): MODE {
      values().forEach {
```

```
      item ->

      if (item.mode == value) {
        return item
      }
    }
    return VIEW
  }
 }
}
```

此处添加了一些附加项。在 enum 的伴生对象中，我们定义了 EXTRAS_KEY，并会在后续操作中对其加以使用，进而理解其功能。除此之外，这里还定义了一个方法，并根据对应值生成 enum。

与 Note 和 Todo 协同工作的两个活动共享同一个类。打开 ItemActivity 并对其进行扩展，如下所示。

```
abstract class ItemActivity : BaseActivity() {
 protected var mode = MODE.VIEW
 override fun getActivityTitle() = R.string.app_name
 override fun onCreate(savedInstanceState: Bundle?) {
   super.onCreate(savedInstanceState)
   val modeToSet = intent.getIntExtra(MODE.EXTRAS_KEY,
   MODE.VIEW.mode)
   mode = MODE.getByValue(modeToSet)
   Log.v(tag, "Mode [ $mode ]")
 }
}
```

这里引入了刚刚定义的类型字段模式，并可通知我们是否正在查看、创建或编辑 Note 或 Todo 条目。随后，代码重载了 onCreate()方法，这一点十分重要。当单击按钮并开启一项活动时，将向其传递多个值。相应地，上述代码片段接收了所传递的值。对此，需要访问 Intent 实例（稍后将对此加以解释）和名为 MODE（MODE.EXTRAS_KEY 的值）的整型字段——getIntExtra()方法即为生成该值的对应方法，且针对每种类型定义了不同的方法版本。如果不存在相应值，则返回 MODE.VIEW.mode。最后，可将该 mode 设置为某个值，而该值通过获取整型值的 MODE 实例得到。

最后一个问题是触发某项活动。对此，打开 ItemsFragment 并按照下列方式对其进行扩展：

```
class ItemsFragment : BaseFragment() {
```

```
...
override fun onCreateView(
  inflater: LayoutInflater?,
  container: ViewGroup?,
  savedInstanceState: Bundle?
): View? {
  val view = inflater?.inflate(getLayout(), container, false)
  val btn = view?.findViewById<FloatingActionButton>
  (R.id.new_item)
  btn?.setOnClickListener {
    val items = arrayOf(
      getString(R.string.todos),
      getString(R.string.notes)
    )
    val builder =
    AlertDialog.Builder(this@ItemsFragment.context)
    .setTitle(R.string.choose_a_type)
    .setItems(
      items,
      { _, which ->
        when (which) {
          0 -> {
            openCreateTodo()
          }
          1 -> {
            openCreateNote()
          }
          else -> Log.e(logTag, "Unknown option selected
          [ $which ]")
        }
      }
    )

    builder.show()
  }

  return view
}

private fun openCreateNote() {
  val intent = Intent(context, NoteActivity::class.java)
  intent.putExtra(MODE.EXTRAS_KEY, MODE.CREATE.mode)
```

```
    startActivity(intent)
  }

  private fun openCreateTodo() {
    val intent = Intent(context, TodoActivity::class.java)
    intent.putExtra(MODE.EXTRAS_KEY, MODE.CREATE.mode)
    startActivity(intent)
  }

}
```

此处访问了 FloatingActionButton 实例，并分配了一个单击监听器。当执行单击操作时，将生成包含两个选项的对话框。其中，每个选项将针对活动开启行为触发相应的方法。另外，两个方法的实现均较为相似。作为示例，我们将重点考查 openCreateNote()方法。

接下来将创建一个新的 Intent 实例。在 Android 中，Intent 代表执行某项任务的意图。当启用某项活动时，需要传递当前上下文和希望启动的活动类。此外，还需要向其中分配相关值，对应值将被传递至活动实例中。在当前示例中，将针对 MODE.CREATE. startActivity()方法传递整型值，进而执行意图并显示最终的屏幕。

运行应用程序，单击屏幕右下方的圆形按钮，并从对话框中完成选项，对应结果如图 4.3 所示。

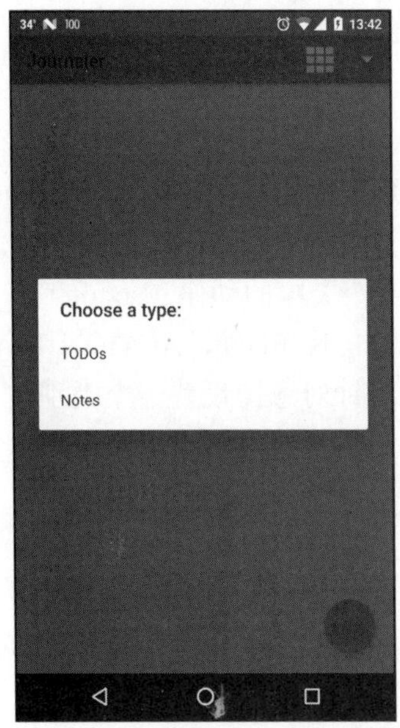

图 4.3

随后将显示如图 4.4 所示的屏幕画面。

接下来，可进一步添加日期和时间等数据，如图 4.5 所示。

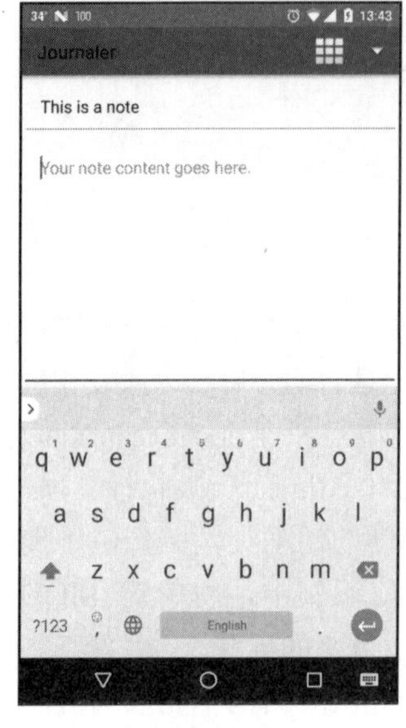

图 4.4

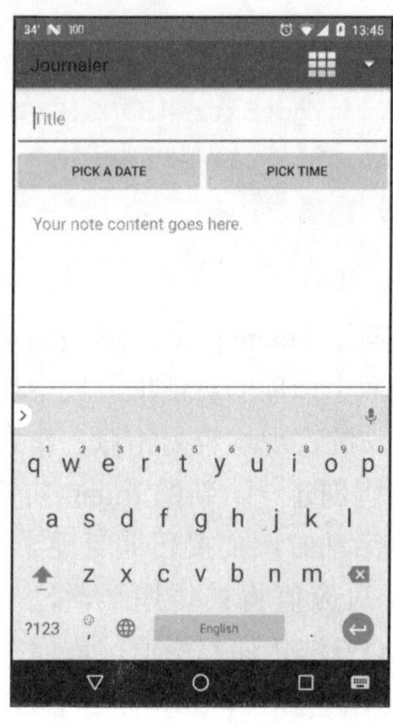

图 4.5

4.4　Android 意图

在 Android 中，大多数计划执行的操作均可通过 Intent 类加以定义。Intent 可用于启动活动、启用服务（后台运行的进程）或者发送广播消息。

Intent 通常会接收传递至某个类中的动作（action）和数据。相应地，可设置的动作属性包括 ACTION_VIEW、ACTION_EDIT、ACTION_MAIN。

除了动作和数据之外，还可针对意图设置一个类别。这里，类别可针对所设置的动作生成附加信息。除此之外，还可针对意图设置类型，以及针对所用的显式组件类名设置组件。

Intent 包含两种类型，即显式意图和隐式意图。

显式意图包含了一个显式组件集，该集合提供了所运行的显式类；隐式意图并不包含显式组件，但系统会根据所分配的数据和属性决定如何对其进行处理。此外，意图解决过程则负责处理此类意图。

下面通过具体示例进一步理解意图的功能。

- 打开一个 Web 页面，如下所示。

```
val intent = Intent(Intent.ACTION_VIEW,
Uri.parse("http://google.com"))
startActivity(intent)
Sharing:
val intent = Intent(Intent.ACTION_SEND)
intent.type = "text/plain"
intent.putExtra(Intent.EXTRA_TEXT, "Check out this cool app!")
startActivity(intent)
```

- 从摄像头中捕捉一幅画面，如下所示。

```
val takePicture = Intent(MediaStore.ACTION_IMAGE_CAPTURE)
if (takePicture.resolveActivity(packageManager) != null) {
  startActivityForResult(takePicture, REQUEST_CAPTURE_PHOTO +
                          position)
} else {
   logger.e(tag, "Can't take picture.")
}
```

- 从图库中选择一幅图像，如下所示。

```
val pickPhoto = Intent(
  Intent.ACTION_PICK,
  MediaStore.Images.Media.EXTERNAL_CONTENT_URI
)
startActivityForResult(pickPhoto, REQUEST_PICK_PHOTO +
                        position)
```

可以看到，Intent 是 Android 框架中的重要组成部分。在 4.5 节中，我们将进一步对代码进行扩展，以充分利用 Intent 这一概念。

4.5　在活动和片段间传递信息

当在活动间传递信息时，需要使用 Android Bundle。Bundle 可包含多个不同类型的值，下面将通过扩展代码展示 Bundle 的使用方式。对此，打开 ItemsFragemnt，并按照下列方式进行更新：

```
private fun openCreateNote() {
 val intent = Intent(context, NoteActivity::class.java)
 val data = Bundle()
```

```
  data.putInt(MODE.EXTRAS_KEY, MODE.CREATE.mode)
  intent.putExtras(data)
  startActivityForResult(intent, NOTE_REQUEST)
}
private fun openCreateTodo() {
  val date = Date(System.currentTimeMillis())
  val dateFormat = SimpleDateFormat("MMM dd YYYY", Locale.ENGLISH)
  val timeFormat = SimpleDateFormat("MM:HH", Locale.ENGLISH)

  val intent = Intent(context, TodoActivity::class.java)
  val data = Bundle()
  data.putInt(MODE.EXTRAS_KEY, MODE.CREATE.mode)
  data.putString(TodoActivity.EXTRA_DATE, dateFormat.format(date))
  data.putString(TodoActivity.EXTRA_TIME,
  timeFormat.format(date))
  intent.putExtras(data)
  startActivityForResult(intent, TODO_REQUEST)
}

override fun onActivityResult(requestCode: Int, resultCode: Int,
data: Intent?) {
  super.onActivityResult(requestCode, resultCode, data)
  when (requestCode) {
    TODO_REQUEST -> {
      if (resultCode == Activity.RESULT_OK) {
        Log.i(logTag, "We created new TODO.")
      } else {
        Log.w(logTag, "We didn't created new TODO.")
      }
    }
    NOTE_REQUEST -> {
      if (resultCode == Activity.RESULT_OK) {
        Log.i(logTag, "We created new note.")
      } else {
        Log.w(logTag, "We didn't created new note.")
        }
    }
  }
}
```

这里引入了一些较为重要的变化内容。首先，需要以子活动方式启用 Note 和 Todo 活动。这意味着，MainActivity 类取决于此类活动的工作结果。当启动子活动（而非

startActivity()方法）时，我们使用了 startActivityForResult()方法。相应地，所传递的参数表示为意图和请求数量。当获得执行结果时，需要重载 onActivityResult()方法。可以看到，我们检查了哪些活动已经完成，以及执行过程是否产生了成功的结果。

除此之外，我们还修改了信息的传递方式。与 Todo 活动类似，此处创建了 Bundle 实例并分配了多个值。同时，还添加了模式、日期和时间内容。相应地，Bundle 通过 putExtras()方法被分配至意图中。当使用这些附加内容时，也可对相关活动进行更新。打开 ItemsActivity 类并按照下列方式添加变化内容：

```
abstract class ItemActivity : BaseActivity() {
 protected var mode = MODE.VIEW
 protected var success = Activity.RESULT_CANCELED
 override fun getActivityTitle() = R.string.app_name

 override fun onCreate(savedInstanceState: Bundle?) {
   super.onCreate(savedInstanceState)
   val data = intent.extras
   data?.let{
     val modeToSet = data.getInt(MODE.EXTRAS_KEY, MODE.VIEW.mode)
     mode = MODE.getByValue(modeToSet)
   }
   Log.v(tag, "Mode [ $mode ]")
 }

 override fun onDestroy() {
   super.onDestroy()
   setResult(success)
 }

}
```

代码中引入了相关字段以加载活动工作的结果。另外，此处还更新了所传递信息的处理方式。可以看到，如果存在某些附加内容，将针对当前模式获得一个整数值。最后，onDestroy()方法还将设置针对父活动可用的工作结果。

打开 TodoActivity 类并按照下列方式添加变化内容：

```
class TodoActivity : ItemActivity() {

companion object {
  val EXTRA_DATE = "EXTRA_DATE"
  val EXTRA_TIME = "EXTRA_TIME"
 }
```

```
  override val tag = "Todo activity"

  override fun getLayout() = R.layout.activity_todo

  override fun onCreate(savedInstanceState: Bundle?) {
    super.onCreate(savedInstanceState)
    val data = intent.extras
    data?.let {
      val date = data.getString(EXTRA_DATE, "")
      val time = data.getString(EXTRA_TIME, "")
      pick_date.text = date
      pick_time.text = time
    }
  }
}
```

至此，我们得到了日期和时间附加值，并将其设置为日期/时间拾取按钮。运行应用程序并打开 Todo 活动，则 Todo 屏幕画面的对应结果如图 4.6 所示。

图 4.6

当退出 Todo 活动并返回主屏幕时，Logcat 中将会显示包含下列内容的日志信息：

```
W/Items fragment--we didn't create a new TODO.
```

由于尚未创建任何 Todo 项，所以此处传递了一个正确的结果。我们通过回到主屏幕取消了创建过程。后续章节还将会继续考查如何成功地创建 Note 和 Todo。

4.6　本章小结

本章讨论了如何连接界面并构建真正的应用程序流。通过向 UI 元素设置相应的动作，本章在屏幕间构建了一个连接，并在点到点间传递数据。虽然可得到正确的结果，但其外观相对简陋。第 5 章将介绍样式化并添加相应的视觉效果。请读者准备好迎接 Android 强大的 UI API 吧！

第 5 章　观　　感

当今，大多数应用程序一般都具有较好的视觉效果，这也使得应用程序更具吸引力。愉悦的体验也能够使得应用程序具有一定的独特性，进而吸引用户安装、使用该程序。本章将讨论如何改善应用程序的外观，并介绍 Android UI 主题方面的内容，同时还将重点阐述 Android 应用程序视觉效果方面的内容。

本章主要涉及以下主题：

- Android 中的主体和颜色样式。
- 与数据资源协同工作。
- 自定义字体和颜色。
- 按钮设计。
- 动画和动画集。

5.1　Android 框架中的主题

第 4 章曾在主 UI 元素间构建了一个连接。如果缺少色彩的支持，应用程序往往难以令人满意。为了获取相应的颜色，下面首先介绍应用程序主题。相应地，我们将扩展现有的某个 Android 主题，并利用喜欢的颜色对其进行重载。

打开 styles.xml 文件，设置应用程序所定义的默认主题并覆写多种颜色。此处将修改 parent 主题，并根据具体需求对其进行定制。下列代码显示了主题的更新操作：

```
<resources>

  <style name="AppTheme"
    parent="Theme.AppCompat.Light.NoActionBar">
    <item name="android:colorPrimary">@color/colorPrimary</item>
    <item name="android:statusBarColor">@color/colorPrimary</item>
    <item name="android:colorPrimaryDark">
     @color/colorPrimaryDark</item>
    <item name="android:colorAccent">@color/colorAccent</item>
    <item name="android:textColor">@android:color/black</item>
  </style>

</resources>
```

代码中定义了一个继承自 AppCompat 的主题。主颜色体现了应用程序自身特有的色彩；而 colorPrimaryDark 则是该颜色的深色变化版本。UI 则控制 colorAccent 中的颜色。除此之外，我们还将主文本颜色设置为黑色；状态栏则采用主颜色。

打开 colors.xml 文件，并定义主题所采用的颜色，如下所示。

```
<?xml version="1.0" encoding="utf-8"?>
<resources>
  <color name="colorPrimary">#ff6600</color>
  <color name="colorPrimaryDark">#197734</color>
  <color name="colorAccent">#ffae00</color>
</resources>
```

在运行应用程序并查看主题效果之前，应确保当前主题已投入使用中。利用下列代码行更新 manifest 文件：

```
<application
android:theme="@style/AppTheme"
```

另外，对于 fragment_items 的浮动按钮，可按照下列方式更新颜色：

```
<android.support.design.widget.FloatingActionButton
    android:backgroundTint="@color/colorPrimary"
    android:id="@+id/new_item"
    android:layout_width="wrap_content"
    android:layout_height="wrap_content"
    android:layout_alignParentBottom="true"
    android:layout_alignParentEnd="true"
    android:layout_margin="@dimen/button_margin" />
```

backgroundTint 属性可确保按钮与状态栏具有相同的颜色。构建并运行当前应用程序，随后可以看到，该应用程序将呈现为橙色效果。

5.2　Android 中的样式

在 5.1 节中定义的主题即体验了某种样式。相应地，全部样式定义于 styles.xml 文件中。随后，本节将创建多种样式，以描述样式的创建方式及其功能。针对按钮、文本或其他视图，均可定义相应的样式。另外，样式还可实现继承操作。

针对样式的功能，可定义一个调色板并用于应用程序中。对此，打开 colors.xml 文件，并按照下列方式对其进行扩展：

```
<color name="green">#11c403</color>
<color name="green_dark">#0e8c05</color>
<color name="white">#ffffff</color>
<color name="white_transparent_40">#64ffffff</color>
<color name="black">#000000</color>
<color name="black_transparent_40">#64000000</color>
<color name="grey_disabled">#d5d5d5</color>
<color name="grey_text">#444d57</color>
<color name="grey_text_transparent_40">#64444d57</color>
<color name="grey_text_middle">#6d6d6d</color>
<color name="grey_text_light">#b9b9b9</color>
<color name="grey_thin_separator">#f1f1f1</color>
<color name="grey_thin_separator_settings">#eeeeee</color>
<color name="vermilion">#f3494c</color>
<color name="vermilion_dark">#c64145</color>
<color name="vermilion_transparent_40">#64f3494c</color>
<color name="plum">#121e2a</color>
```

这里应注意透明色。下面考查白色示例：对于纯白色，对应代码为#ffffff；而包含 40%透明度的白色其代码为#64ffffff。当实现透明度时，可尝试使用下列值。

注意：

0% = #00。

10% = #16。

20% = #32。

30% = #48。

40% = #64。

50% = #80。

60% = #96。

70% = #112。

80% = #128。

90% = #144。

在定义了调色板后，接下来将生成第一种样式。打开 styles.xml 文件并对其进行扩展，如下所示。

```
<style name="simple_button">
  <item name="android:textSize">16sp</item>
  <item name="android:textAllCaps">false</item>
```

```
  <item name="android:textColor">@color/white</item>
</style>

<style name="simple_button_green" parent="simple_button">
  <item name="android:background">
  @drawable/selector_button_green</item>
</style>
```

此处定义了两种样式。其中，第一种样式定义了一个简单的按钮，并包含了一个白色的文本，对应字体的尺寸为 16sp；第二种样式对第一种样式进行了扩展，并添加了背景属性。稍后将创建一个选择器，以便展示样式的定义方式。考虑到当前尚未包含这一类资源，因而可在 drawable resource 文件夹中创建一个 selector_button_green.xml 文件，如下所示。

```
<?xml version="1.0" encoding="utf-8"?>
<selector xmlns:android=
  "http://schemas.android.com/apk/res/android">

  <item android:drawable="@color/grey_disabled"
    android:state_enabled="false" />
  <item android:drawable="@color/green_dark"
    android:state_selected="true" />
  <item android:drawable="@color/green_dark"
    android:state_pressed="true" />
  <item android:drawable="@color/green" />
</selector>
```

这里定义了一个选择器。选择器定义为一个 XML 文件，可描述视觉行为或不同的状态。对于按钮的禁用状态，代码中添加了不同的颜色；而按钮被按下、释放或未执行任何交互操作时，则显示另一种颜色。

当查看按钮的外观时，可打开 activity_todo，并针对每个按钮设置下列样式：

```
style="@style/simple_button_green"
```

随后，运行应用程序并打开 Todo 屏幕，如图 5.1 所示。

当按下按钮时可以看到，颜色将变为深绿色。在 5.4 节中，将通过添加圆角进一步改善按钮的外观。首先创建一些基础样式，如下所示。

- 针对输入框和导航抽屉，将样式添加至 styles.xml 中。

```
<style name="simple_button_grey" parent="simple_button">
 <item name="android:background">
 @drawable/selector_button_grey</item>
```

```
</style>

<style name="edit_text_transparent">
  <item name="android:textSize">14sp</item>
  <item name="android:padding">19dp</item>
  <item name="android:textColor">@color/white</item>
  <item name="android:textColorHint">@color/white</item>
  <item name="android:background">
  @color/black_transparent_40</item>
</style>

<style name="edit_text_gery_text"
  parent="edit_text_transparent">
  <item name="android:textAlignment">textStart</item>
  <item name="android:textColor">@color/white</item>
  <item name="android:background">@color/grey_text_light</item>
</style>
```

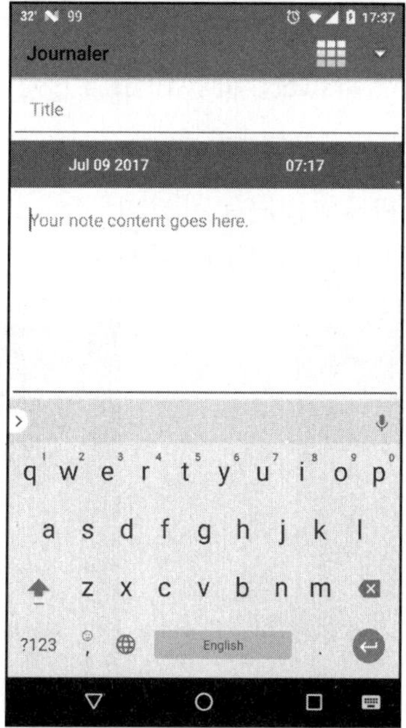

图 5.1

- 针对输入框，此处定义了提示信息的颜色。此外，我们还引入了一个抽屉式的选择器 selector_button_grey。

```
<?xml version="1.0" encoding="utf-8"?>
```

```
<selector xmlns:android=
  "http://schemas.android.com/apk/res/android">

 <item android:drawable="@color/grey_disabled"
  android:state_enabled="false" />
 <item android:drawable="@color/grey_text_middle"
  android:state_selected="true" />
 <item android:drawable="@color/grey_text_middle"
  android:state_pressed="true" />
 <item android:drawable="@color/grey_text" />
</selector>
```

- 对于两个屏幕（Note 和 Todo）上的 note_title，添加下列样式：

```
style="@style/edit_text_transparent"
```

- 对于 note_content，添加下列样式：

```
style="@style/edit_text_gery_text"
```

- 对于 adapter_navigation_drawer，向按钮添加下列样式：

```
style="@style/simple_button_grey"
```

运行应用程序，全部屏幕和导航抽屉如图 5.2 所示。

不难发现，UI 外观已经有所改善。下面查看图 5.3。

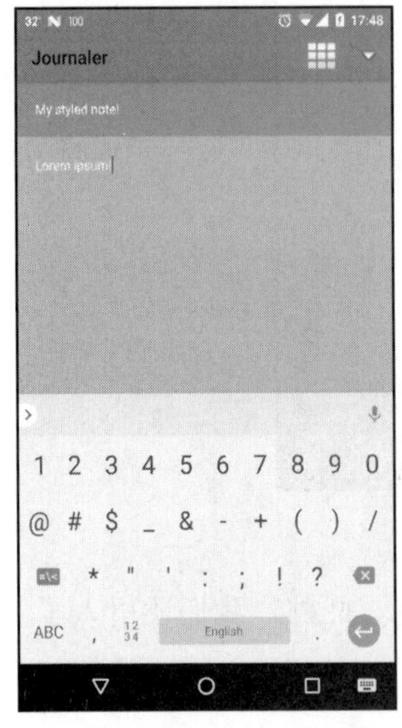

图 5.2

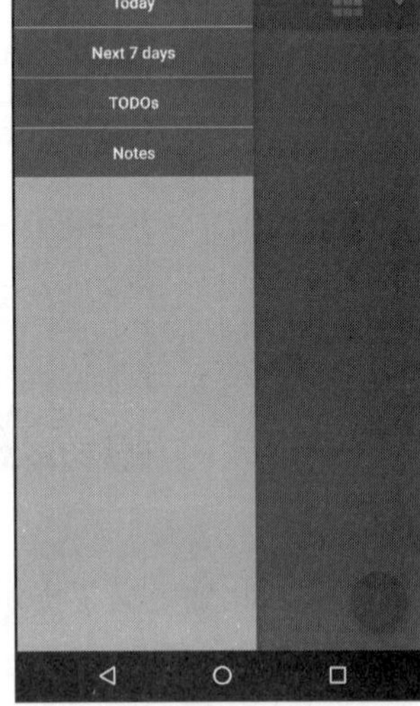

图 5.3

读者可根据具体需求随意调整属性和颜色。但设计工作远未结束，接下来还将对字体予以考查。

5.2.1 与数据资源协同工作

应用程序还可与原始资源协同工作，字体便是其中之一。具体来说，每个字体应用都是一个存储于 assets 文件夹中的独立文件。assets 文件夹是 main 目录的一个子目录，或者是体现构建变化版本的一个目录。除了字体之外，还可存储文本文件、mp3、waw、mid 等。注意，这一类文件类型不可存储于 res 目录中。

5.2.2 使用自定义字体

字体可视为一类数据资源。当为应用程序提供相关字体时，首先需要对其进行复制。网络上涵盖了大量的免费字体资源，例如 Google Fonts。对此，可下载字体并将其复制至 assets 目录中。当前，我们的字体将置于 assets/fonts 目录中。

在当前示例中，我们将使用 Exo。Exo 附带了以下字体：

- ❑ Exo2-Black.ttf。
- ❑ Exo2-BlackItalic.ttf。
- ❑ Exo2-Bold.ttf。
- ❑ Exo2-BoldItalic.ttf。
- ❑ Exo2-ExtraBold.ttf。
- ❑ Exo2-ExtraBoldItalic.ttf。
- ❑ Exo2-ExtraLight.ttf。
- ❑ Exo2-ExtraLightItalic.ttf。
- ❑ Exo2-Italic.ttf。
- ❑ Exo2-Light.ttf。
- ❑ Exo2-LightItalic.ttf。
- ❑ Exo2-Medium.ttf。
- ❑ Exo2-MediumItalic.ttf。
- ❑ Exo2-Regular.ttf。
- ❑ Exo2-SemiBold.ttf。
- ❑ Exo2-SemiBoldItalic.ttf。
- ❑ Exo2-Thin.ttf。

- Exo2-ThinItalic.ttf。

仅将 font 文件复制至 assets 目录中尚不支持字体的使用，还需要通过代码方式应用字体功能。

打开 BaseActivity 并对其进行扩展，如下所示。

```
abstract class BaseActivity : AppCompatActivity() {
 companion object {
   private var fontExoBold: Typeface? = null
   private var fontExoRegular: Typeface? = null

   fun applyFonts(view: View, ctx: Context) {
   var vTag = ""
   if (view.tag is String) {
     vTag = view.tag as String
   }
   when (view) {
     is ViewGroup -> {
       for (x in 0..view.childCount - 1) {
         applyFonts(view.getChildAt(x), ctx)
       }
     }
     is Button -> {
       when (vTag) {
         ctx.getString(R.string.tag_font_bold) -> {
           view.typeface = fontExoBold
         }
         else -> {
           view.typeface = fontExoRegular
         }
       }
     }
     is TextView -> {
       when (vTag) {
         ctx.getString(R.string.tag_font_bold) -> {
           view.typeface = fontExoBold
         }
         else -> {
           view.typeface = fontExoRegular
         }
       }
     }
     is EditText -> {
```

```
      when (vTag) {
        ctx.getString(R.string.tag_font_bold) -> {
          view.typeface = fontExoBold
        }
        else -> {
          view.typeface = fontExoRegular
        }
      }
    }
  }
 }
}
...
override fun onPostCreate(savedInstanceState: Bundle?) {
   super.onPostCreate(savedInstanceState)
   Log.v(tag, "[ ON POST CREATE ]")
   applyFonts()
}
...
protected fun applyFonts() {
   initFonts()
   Log.v(tag, "Applying fonts [ START ]")
   val rootView = findViewById(android.R.id.content)
   applyFonts(rootView, this)
   Log.v(tag, "Applying fonts [ END ]")
}

private fun initFonts() {
   if (fontExoBold == null) {
     Log.v(tag, "Initializing font [ Exo2-Bold ]")
     fontExoBold = Typeface.createFromAsset(assets, "fonts/Exo2-
     Bold.ttf")
   }
   if (fontExoRegular == null) {
     Log.v(tag, "Initializing font [ Exo2-Regular ]")
     fontExoRegular = Typeface.createFromAsset(assets,
     "fonts/Exo2-Regular.ttf")
   }
 }
}
```

此处扩展了基本活动以对字体进行处理。当活动进入 onPostCreate()时，applyFonts()方法将被调用。applyFonts()方法将执行下列操作：

- 调用 initFonts()方法，该方法将从数据资源中创建 TypeFace 实例。TypeFace 用于字体的表达及其视觉属性。这里将针对 ExoBold 和 ExoRegular 实例化 TypeFace。
- 接下来将针对当前活动获取 root 视图，并将其传递至源自伴生对象的 applyFonts()方法中。如果视图是一个 view group，则遍历其子元素，直至到达普通的视图。视图包含了一个 typeface 属性，进而设置 typeface 实例。此外，还需要检索每个视图中的名为 tag 类属性。在 Android 中，可针对视图设置 tag。这里，tag（标签）可以是任何类的实例。在当前示例中，将检查 tag 是否为 String（基于名为 tag_font_bold 的字符串资源值）。

当设置标签时，可在 values 目录中创建一个名为 tags 的新的 xml 文件并使用下列内容填充：

```
<?xml version="1.0" encoding="utf-8"?>
<resources>
  <string name="tag_font_regular">FONT_REGULAR</string>
  <string name="tag_font_bold">FONT_BOLD</string>
</resources>
To apply it open styles.xml and add tag to simple_button style:
<item name="android:tag">@string/tag_font_bold</item>
```

随后，应用程序的按钮将包含粗体字。构建并运行应用程序，可以看到，字体已发生了变化。

5.3　应用颜色

前述内容曾对应用程序定义了调色板，通过访问其中的资源，我们即可使用到各种颜色。但有些时候，我们没有一种特定的颜色资源可用，则可通过后端（在某些 API 调用的响应结果中）并以动态方式获取相关颜色，或者出于某些原因需要在代码中定义颜色。

当需要在代码中处理颜色问题时，Android 对此提供了强力的支持，本节将通过一些示例阐述具体的操作方式。

当从现有的资源中获取颜色时，可执行下列操作：

```
val color = ContextCompat.getColor(contex, R.color.plum)
```

之前的操作方式如下：

```
val color = resources.getColor(R.color.plum)
```

上述操作方式已在 Android 版本 6 中被弃用。

在获得相关颜色后，可在某个视图上对其加以使用，如下所示。

```
pick_date.setTextColor(color)
```

另一种获取颜色的方法是访问 Color 类的静态方法。下列将首先解析颜色字符串：

```
val color = Color.parseColor("#ff0000")
```

注意，某些颜色已被预定义：

```
val color = Color.RED
```

因而无须解析#ff0000。类似地，其他颜色还包括：

```
public static final int BLACK
public static final int BLUE
public static final int CYAN
public static final int DKGRAY
public static final int GRAY
public static final int GREEN
public static final int LTGRAY
public static final int MAGENTA
public static final int RED
public static final int TRANSPARENT
public static final int WHITE
public static final int YELLOW
```

有时，可能仅需要使用与红、绿、蓝相关的参数，并在此基础上创建某种颜色，如下所示。

```
Color red = Color.valueOf(1.0f, 0.0f, 0.0f);
```

注意，上述方法在版本 26 的 API 中已得到了支持。

如果 RGB 并不是所期望的颜色空间，则可将其作为参数予以传递，如下所示。

```
val colorSpace = ColorSpace.get(ColorSpace.Named.NTSC_1953)
val color = Color.valueOf(1f, 1f, 1f, 1f, colorSpace)
```

可以看到，当处理颜色问题时，存在多种可能性。如果标准的颜色资源无法满足当前要求，还可采用更加高级的方式对其加以处理。这里也鼓励用户进行多方尝试。

例如，如果使用 AppCompat 库，那么，一旦获得了 Color 实例，则可按照下列方式对其加以使用：

```
counter.setTextColor(
  ContextCompat.getColor(context, R.color.vermilion)
)
```

下面考查图 5.4。

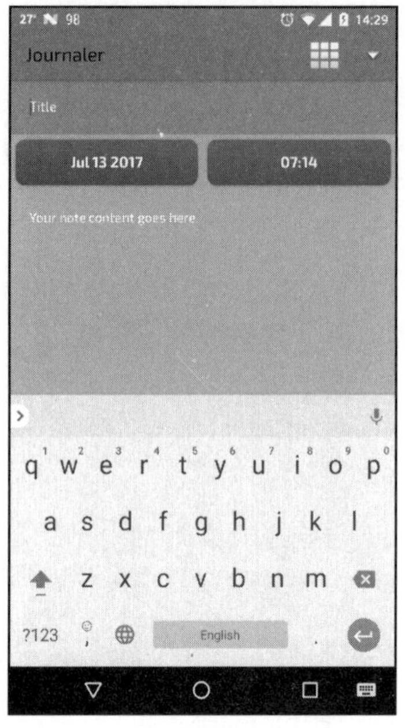

图 5.4

5.4　改进按钮的外观

之前曾对按钮提供了相应的颜色并为其定义了各种状态；同时，还采用了不同的方式对每种状态设置了颜色。具体来说，我们针对禁用状态、启用状态和按下状态指定了不同的颜色，接下来将进一步对此加以考查。本节将生成圆角按钮，并尝试采用渐变颜色（而非固定颜色）。此外，还将对新按钮样式设置布局。对此，打开 activity_todo 布局，并调整按钮容器，如下所示。

```
<LinearLayout
  android:background="@color/grey_text_light"
  android:layout_width="match_parent"
  android:layout_height="wrap_content"
  android:orientation="horizontal"
  android:weightSum="1">
  ...

</LinearLayout>
```

这里设置了与编辑文本框相同的背景。另外，按钮将呈现为圆角状，并于其他组件位于同一背景中。接下来将定义一些附加尺寸和颜色，其中包括圆角按钮的半径，如下所示。

```
<dimen name="button_corner">10dp</dimen>
```

考虑到将使用渐变颜色，因而需要针对这一特征添加第二种颜色。相应地，可向 colors.xml 中加入此类颜色，如下所示。

```
<color name="green2">#208c18</color>
<color name="green_dark2">#0b5505</color>
```

当完成上述定义时，需要针对绿色按钮更新当前样式，如下所示。

```
<style name="simple_button_green" parent="simple_button">
  <item name="android:layout_margin">5dp</item>
  <item name="android:background">
  @drawable/selector_button_green</item>
</style>
```

鉴于设置了边距，因而按钮键彼此间处于分离状态。下面将创建矩形圆角可绘制对象，并生成以下 3 种资源：rect_rounded_green、rect_rounded_green_dark 和 rect_rounded_grey_disabled，其定义方式如下所示。

- rect_rounded_green：

```
<shape xmlns:android=
 "http://schemas.android.com/apk/res/android">
  <gradient
  android:angle="270"
  android:endColor="@color/green2"
  android:startColor="@color/green" />

 <corners android:radius="@dimen/button_corner" />
</shape>
```

- rect_rounded_green_dark：

```
<shape xmlns:android="http://schemas.android.com/apk/res/android">
 <gradient
 android:angle="270"
 android:endColor="@color/green_dark2"
 android:startColor="@color/green_dark" />

 <corners android:radius="@dimen/button_corner" />
```

```
</shape>
```

- rect_rounded_grey_disabled：

```
<shape xmlns:android=
"http://schemas.android.com/apk/res/android">

<solid android:color="@color/grey_disabled" />
<corners android:radius="@dimen/button_corner" />
</shape>
```

- 针对下列属性定义渐变色：
  - 渐变角（270°）。
  - 起始颜色（使用相应的颜色资源）。
  - 结束颜色（使用相应的颜色资源）。

另外，每个可绘制资源均包含了一个圆角半径值。最后一步是更新选择器，对此，打开 selector_button_green，并按照下列方式进行更新：

```
<?xml version="1.0" encoding="utf-8"?>
<selector xmlns:android=
  "http://schemas.android.com/apk/res/android">

<item
  android:drawable="@drawable/rect_rounded_grey_disabled"
  android:state_enabled="false" />

<item
  android:drawable="@drawable/rect_rounded_green_dark"
  android:state_selected="true" />

<item
  android:drawable="@drawable/rect_rounded_green_dark"
  android:state_pressed="true" />

<item
  android:drawable="@drawable/rect_rounded_green" />

</selector>
```

构建并运行应用程序。打开 Todo 屏幕并查看相关结果。当前，按钮包含了圆角且外观得到了进一步的改善。由于边距的存在，按钮彼此间处于分离状态。当按下按钮时，将会看到基于所定义的深绿色的第二种渐变色，如图 5.5 所示。

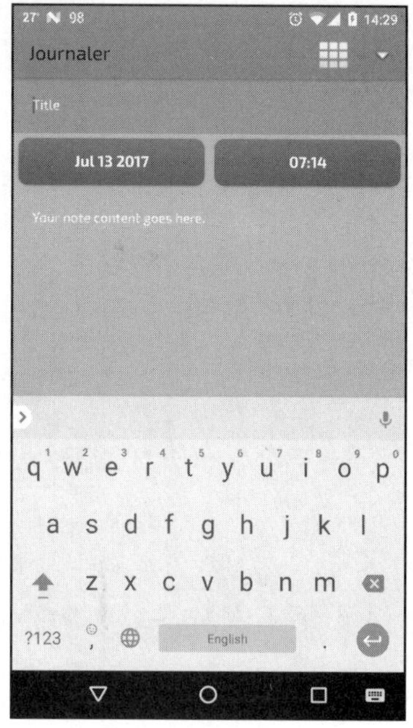

图 5.5

5.5 设置动画

如果希望当前布局效果更具娱乐性，同时丰富用户的使用体验，可尝试向其添加动画效果。动画效果可通过代码或属性的动画行为加以定义。本节将添加简单、有效的开启动画，进而改善每个屏幕。

动画也定义为一种资源，并位于 anim 资源目录中。此处需要使用以下几种动画资源，即 fade_in、fade_out、bottom_to_top、top_to_bottom、hide_to_top、hide_to_bottom，其创建和定义方式如下所示。

- fade_in：

```
<?xml version="1.0" encoding="utf-8"?>
<alpha xmlns:android=
"http://schemas.android.com/apk/res/android"
android:duration="300"
android:fromAlpha="0.0"
android:interpolator="@android:anim/accelerate_interpolator"
android:toAlpha="1.0" />
```

- fade_out：

```
<?xml version="1.0" encoding="utf-8"?>
<alpha xmlns:android=
  "http://schemas.android.com/apk/res/android"
  android:duration="300"
  android:fillAfter="true"
  android:fromAlpha="1.0"
  android:interpolator="@android:anim/accelerate_interpolator"
  android:toAlpha="0.0" />
-  bottom_to_top:
<set xmlns:android=
  "http://schemas.android.com/apk/res/android"
  android:fillAfter="true"
  android:fillEnabled="true"
  android:shareInterpolator="false">

<translate
  android:duration="900"
  android:fromXDelta="0%"
  android:fromYDelta="100%"
  android:toXDelta="0%"
  android:toYDelta="0%" />

</set>
```

- top_to_bottom：

```
<set xmlns:android="http://schemas.android.com/apk/res/android"
  android:fillAfter="true"
  android:fillEnabled="true"
  android:shareInterpolator="false">

<translate
  android:duration="900"
  android:fromXDelta="0%"
  android:fromYDelta="-100%"
  android:toXDelta="0%"
  android:toYDelta="0%" />
</set>
```

- hide_to_top：

```
<set xmlns:android="http://schemas.android.com/apk/res/android"
```

```
  android:fillAfter="true"
  android:fillEnabled="true"
  android:shareInterpolator="false">

<translate
  android:duration="900"
  android:fromXDelta="0%"
  android:fromYDelta="0%"
  android:toXDelta="0%"
  android:toYDelta="-100%" />

</set>
```

- hide_to_bottom：

```
<set xmlns:android=
  "http://schemas.android.com/apk/res/android"
  android:fillAfter="true"
  android:fillEnabled="true"
  android:shareInterpolator="false">

<translate
  android:duration="900"
  android:fromXDelta="0%"
  android:fromYDelta="0%"
  android:toXDelta="0%"
  android:toYDelta="100%" />

</set>
```

在褪色动画中，针对视图定义了一个 alpha 属性，同时还设置了动画时长、alpha 值、用于动画的插值器。在 Android 中，可选择以下几种插值器之一供动画使用：

- accelerate_interpolator。
- accelerate_decelerate_interpolator。
- bounce_interpolator。
- cycle_interpolator。
- anticipate_interpolator。
- anticipate_overshot_interpolator。
- 其他插值器定义于@android:anim/...中。

另外，还可通过 from 和 to 参数定义平移动画。

在使用动画之前，还需要对背景进行适当调整，以使在动画开始前的布局中不存在任何缝隙。对于 activity_main，可针对 Toolbar parent 视图添加背景，如下所示。

```
android:background="@android:color/darker_gray"
```

对于另一个父工具栏中的 activity_note 和 activity_todo 嵌套工具栏，最终颜色与工具栏下方标题栏的颜色相同，如下所示。

```
<LinearLayout
  android:layout_width="match_parent"
  android:layout_height="wrap_content"
  android:background="@color/black_transparent_40"
  android:orientation="vertical">

<LinearLayout
  android:layout_width="match_parent"
  android:layout_height="wrap_content"
  android:background="@color/black_transparent_40"
  android:orientation="vertical">

<android.support.v7.widget.Toolbar
  android:id="@+id/toolbar"
  android:layout_width="match_parent"
  android:layout_height="50dp"
  android:background="@color/colorPrimary"
  android:elevation="4dp" />
```

最终，针对各个屏幕实现淡入淡出效果。对此，打开 BaseActivity，并按照下列方式进行调整：

```
override fun onCreate(savedInstanceState: Bundle?) {
  super.onCreate(savedInstanceState)
  overridePendingTransition(R.anim.fade_in, R.anim.fade_out)
  setContentView(getLayout())
  setSupportActionBar(toolbar)
  Log.v(tag, "[ ON CREATE ]")
}
```

此处通过 overridePendingTransition()方法重载了渐变效果，该方法接收淡入和淡出参数。除此之外，还需要更新 onResume()方法和 onPause()方法，如下所示。

```
override fun onResume() {
  super.onResume()
  Log.v(tag, "[ ON RESUME ]")
```

```
    val animation = getAnimation(R.anim.top_to_bottom)
    findViewById(R.id.toolbar).startAnimation(animation)
}

override fun onPause() {
    super.onPause()
    Log.v(tag, "[ ON PAUSE ]")
    val animation = getAnimation(R.anim.hide_to_top)
    findViewById(R.id.toolbar).startAnimation(animation)

}
```

代码中创建了动画实例，并通过 startAnimation()方法将其应用于视图上；getAnimation()方法则是之前所定义的方法，因而可向 BaseActivity 添加其实现，如下所示。

```
protected fun getAnimation(animation: Int): Animation =
AnimationUtils.loadAnimation(this, animation)
```

鉴于当前采用了 Kotlin 语言，因而应使其对所有的活动均为可用，而不仅仅是那些扩展了 BaseActivity 方法的活动，如下所示。

```
fun Activity.getAnimation(animation: Int): Animation =
AnimationUtils.loadAnimation(this, animation)
```

再次构建、运行应用程序。多次打开、关闭应用程序以查看动画的工作方式。

5.6 Android 中的动画集

在前述章节中，我们曾与定义于 XML 文件中的资源动画协同工作。本节将尝试利用视图属性和动画集，并通过简单、有效的示例展示动画的功能和应用方式。

下面通过代码描述第一个动画。对此，打开 ItemsFragment，并添加下列方法：

```
private fun animate(btn: FloatingActionButton, expand: Boolean =
true) {
   btn.animate()
           .setInterpolator(BounceInterpolator())
           .scaleX(if(expand){ 1.5f } else { 1.0f })
           .scaleY(if(expand){ 1.5f } else { 1.0f })
           .setDuration(2000)
           .start()
}
```

上述方法通过插值方式实现了按钮的缩放动画。如果 expand 参数为 true，则按钮处于放大状态；否则，按钮将处于缩小状态。

随后，将其应用于浮动按钮上，并扩展按钮的单击监听器，如下所示。

```
btn?.setOnClickListener {
  animate(btn)
  ...

}
```

主对话框的设置方式如下所示（可撤销性）。

```
val builder = AlertDialog.Builder(this@ItemsFragment.context)
                .setTitle(R.string.choose_a_type)
                .setCancelable(true)
                .setOnCancelListener {
                    animate(btn, false)
                }
.setItems( ... )

...
builder.show()
```

构建和运行应用程序。单击 add item 按钮，并于随后单击外部区域关闭对话框。其间，我们可以看到相应的缩放动画。

在浮动按钮的完成阶段，还需要向“+”号添加 PNG 资源，并将其应用于该按钮上，如下所示。

```
<android.support.design.widget.FloatingActionButton
...
android:src="@drawable/add"
android:scaleType="centerInside"
...
/>
```

在向浮动按钮添加了图标后，动画外观得到了进一步的改善。下面创建包含多个动画的动画集，如下所示。

```
private fun animate(btn: FloatingActionButton, expand: Boolean =
true) {
   val animation1 = ObjectAnimator.ofFloat(btn, "scaleX",
   if(expand){ 1.5f } else { 1.0f })
   animation1.duration = 2000
```

```
    animation1.interpolator = BounceInterpolator()

    val animation2 = ObjectAnimator.ofFloat(btn, "scaleY",
    if(expand){ 1.5f } else { 1.0f })
    animation2.duration = 2000
    animation2.interpolator = BounceInterpolator()

    val animation3 = ObjectAnimator.ofFloat(btn, "alpha",
    if(expand){ 0.3f } else { 1.0f })
    animation3.duration = 500
    animation3.interpolator = AccelerateInterpolator()

    val set = AnimatorSet()
    set.play(animation1).with(animation2).before(animation3)
    set.start()
}
```

AnimatorSet 类可生成相对复杂的动画。在当前示例中，分别针对 x 轴和 y 轴定义了缩放动画，并以同步方式生成动画，进而在两个方向上实现了缩放效果。在视图被缩放后，视图空间将被减少（或增加）。

构建并运行当前项目，随后即可看到新呈现的动画效果。

5.7　本 章 小 结

本章内容颇具交互特性。首先，本章介绍了如何添加、定义、修改和调整 Android 中的主题，随后则讨论了 Android 中的样式和数据资源；其间，我们还使用到了自定义字体和颜色。最后，本章还制作了具有优美外观的按钮，并实现了一些简单的动画效果。第 6 章将开始介绍 Android Framework 的系统部分，并将首先探讨权限问题。

第 6 章　权　　限

在介绍了用户界面后，本章将步入本书中较为复杂的部分——系统。

本章和后续章节将深入分析 Android 系统的结构，其中涉及权限、数据库处理、偏好设置、并发设计、服务、消息传输、后端操作、API 以及性能方面的问题。

然而，本书并不打算介绍 Android 框架中的全部内容，这超出了本书讨论范围。Android 是一类大型框架，完全掌握其中的内容可能会需要好几年的时间。因此，本书仅涉及与 Android 和 Kotli 相关的知识。

本章将探讨 Android 权限方面的问题，包括应用场合以及使用原因（后者更应引起读者的足够重视）。

本章主要涉及以下主题：

- ❑ Android Manifest 中的权限。
- ❑ 请求权限。
- ❑ 采用 Kotlin 方式的权限处理。

6.1　Android Manifest 中的权限

Android 应用程序在其自身的进程中运行，且与操作系统的其他部分隔离。据此，当执行某些特定于系统的操作时，需要对权限予以请求。此类权限请求的例子包括使用蓝牙设备、检索当前 GPS 位置、发送 SMS 消息、读/写文件系统。对此，存在多种方式可处理权限问题。下面讨论较为基本的 manifest。

首先，需要确定需要哪些权限。可能存在的情况是，在安装过程中，用户决定不安装应用程序（可能存在太多的授权）。例如，当应用程序本身只是一个简单的图片库应用程序时，用户可能会询问为什么应用程序需要发送 SMS 功能。

对于本章所要开发的 Journaler 应用程序来说，将会使用下列权限：

- ❑ 读取 GPS 坐标，因为我们希望所创建的每个音符都具有关联的坐标。
- ❑ 可以访问互联网，并于随后执行 API 调用。
- ❑ 启动完成事件，这样应用程序服务就可以在每次重新启动手机时与后端同步。
- ❑ 读/写外部存储，进而可读取或存储数据。
- ❑ 访问网络状态，进而查看互联网连接状态。

- 利用振动功能作为接收提示。

打开 AndroidManifest.xml 文件，并通过下列权限进行更新：

```
<manifest xmlns:android=
 "http://schemas.android.com/apk/res/android"
 package="com.journaler">

   <uses-permission android:name="android.permission.INTERNET" />
   <uses-permission android:name=
     "android.permission.RECEIVE_BOOT_COMPLETED" />
   <uses-permission android:name=
     "android.permission.READ_EXTERNAL_STORAGE" />
   <uses-permission android:name=
     "android.permission.WRITE_EXTERNAL_STORAGE" />
   <uses-permission android:name=
     "android.permission.ACCESS_NETWORK_STATE" />
   <uses-permission android:name=
     "android.permission.ACCESS_FINE_LOCATION" />
   <uses-permission android:name=
     "android.permission.ACCESS_COARSE_LOCATION" />
   <uses-permission android:name="android.permission.VIBRATE" />
     <application ... >
       ...
     </application>

     ...
</manifest>
```

这里，所请求的权限名称非常容易理解，它们涵盖了我们提到的所有要点。除了上述权限之外，还可以请求其他一些权限，如下所示。

```
<uses-permission android:name=
"android.permission.ACCESS_CHECKIN_PROPERTIES" />
<uses-permission android:name=
"android.permission.ACCESS_LOCATION_EXTRA_COMMANDS" />
<uses-permission android:name=
"android.permission.ACCESS_MOCK_LOCATION" />
<uses-permission android:name=
"android.permission.ACCESS_SURFACE_FLINGER" />
<uses-permission android:name=
"android.permission.ACCESS_WIFI_STATE" />
<uses-permission android:name=
```

```
"android.permission.ACCOUNT_MANAGER" />
<uses-permission android:name=
"android.permission.AUTHENTICATE_ACCOUNTS" />
<uses-permission android:name=
"android.permission.BATTERY_STATS" />
<uses-permission android:name=
"android.permission.BIND_APPWIDGET" />
<uses-permission android:name=
"android.permission.BIND_DEVICE_ADMIN" />
<uses-permission android:name=
"android.permission.BIND_INPUT_METHOD" />
<uses-permission android:name=
"android.permission.BIND_REMOTEVIEWS" />
<uses-permission android:name=
"android.permission.BIND_WALLPAPER" />
<uses-permission android:name=
"android.permission.BLUETOOTH" />
<uses-permission android:name=
"android.permission.BLUETOOTH_ADMIN" />
<uses-permission android:name=
"android.permission.BRICK" />
<uses-permission android:name=
"android.permission.BROADCAST_PACKAGE_REMOVED" />
<uses-permission android:name=
"android.permission.BROADCAST_SMS" />
<uses-permission android:name=
"android.permission.BROADCAST_STICKY" />
<uses-permission android:name=
"android.permission.BROADCAST_WAP_PUSH" />
<uses-permission android:name=
"android.permission.CALL_PHONE"/>
<uses-permission android:name=
"android.permission.CALL_PRIVILEGED" />
<uses-permission android:name=
"android.permission.CAMERA"/>
<uses-permission android:name=
"android.permission.CHANGE_COMPONENT_ENABLED_STATE" />
<uses-permission android:name=
"android.permission.CHANGE_CONFIGURATION" />
<uses-permission android:name=
"android.permission.CHANGE_NETWORK_STATE" />
<uses-permission android:name=
"android.permission.CHANGE_WIFI_MULTICAST_STATE" />
<uses-permission android:name=
```

```
"android.permission.CHANGE_WIFI_STATE" />
<uses-permission android:name=
"android.permission.CLEAR_APP_CACHE" />
<uses-permission android:name=
"android.permission.CLEAR_APP_USER_DATA" />
<uses-permission android:name=
"android.permission.CONTROL_LOCATION_UPDATES" />
<uses-permission android:name=
"android.permission.DELETE_CACHE_FILES" />
<uses-permission android:name=
"android.permission.DELETE_PACKAGES" />
<uses-permission android:name=
"android.permission.DEVICE_POWER" />
<uses-permission android:name=
"android.permission.DIAGNOSTIC" />
<uses-permission android:name=
"android.permission.DISABLE_KEYGUARD" />
<uses-permission android:name=
"android.permission.DUMP" />
<uses-permission android:name=
"android.permission.EXPAND_STATUS_BAR" />
<uses-permission android:name="
android.permission.FACTORY_TEST" />
<uses-permission android:name=
"android.permission.FLASHLIGHT" />
<uses-permission android:name=
"android.permission.FORCE_BACK" />
<uses-permission android:name=
"android.permission.GET_ACCOUNTS" />
<uses-permission android:name=
"android.permission.GET_PACKAGE_SIZE" />
<uses-permission android:name=
"android.permission.GET_TASKS" />
<uses-permission android:name=
"android.permission.GLOBAL_SEARCH" />
<uses-permission android:name=
"android.permission.HARDWARE_TEST" />
<uses-permission android:name=
"android.permission.INJECT_EVENTS" />
<uses-permission android:name=
"android.permission.INSTALL_LOCATION_PROVIDER" />
<uses-permission android:name=
"android.permission.INSTALL_PACKAGES" />
<uses-permission android:name=
```

```
"android.permission.INTERNAL_SYSTEM_WINDOW" />
<uses-permission android:name=
"android.permission.KILL_BACKGROUND_PROCESSES" />
<uses-permission android:name=
"android.permission.MANAGE_ACCOUNTS" />
<uses-permission android:name=
"android.permission.MANAGE_APP_TOKENS" />
<uses-permission android:name=
"android.permission.MASTER_CLEAR" />
<uses-permission android:name=
"android.permission.MODIFY_AUDIO_SETTINGS" />
<uses-permission android:name=
"android.permission.MODIFY_PHONE_STATE" />
<uses-permission android:name=
"android.permission.MOUNT_FORMAT_FILESYSTEMS" />
<uses-permission android:name=
"android.permission.MOUNT_UNMOUNT_FILESYSTEMS" />
<uses-permission android:name=
"android.permission.NFC" />
<uses-permission android:name=
"android.permission.PROCESS_OUTGOING_CALLS" />
<uses-permission android:name=
"android.permission.READ_CALENDAR" />
<uses-permission android:name=
"android.permission.READ_CONTACTS" />
<uses-permission android:name=
"android.permission.READ_FRAME_BUFFER" />
<uses-permission android:name=
"android.permission.READ_HISTORY_BOOKMARKS" />
<uses-permission android:name=
"android.permission.READ_INPUT_STATE" />
<uses-permission android:name=
"android.permission.READ_LOGS" />
<uses-permission android:name=
"android.permission.READ_PHONE_STATE" />
<uses-permission android:name=
"android.permission.READ_SMS" />
<uses-permission android:name=
"android.permission.READ_SYNC_SETTINGS" />
<uses-permission android:name=
"android.permission.READ_SYNC_STATS" />
<uses-permission android:name=
"android.permission.REBOOT" />
<uses-permission android:name=
```

```
"android.permission.RECEIVE_MMS" />
<uses-permission android:name=
"android.permission.RECEIVE_SMS" />
<uses-permission android:name=
"android.permission.RECEIVE_WAP_PUSH" />
<uses-permission android:name=
"android.permission.RECORD_AUDIO" />
<uses-permission android:name=
"android.permission.REORDER_TASKS" />
<uses-permission android:name=
"android.permission.RESTART_PACKAGES" />
<uses-permission android:name=
"android.permission.SEND_SMS" />
<uses-permission android:name=
"android.permission.SET_ACTIVITY_WATCHER" />
<uses-permission android:name=
"android.permission.SET_ALARM" />
<uses-permission android:name=
"android.permission.SET_ALWAYS_FINISH" />
<uses-permission android:name=
"android.permission.SET_ANIMATION_SCALE" />
<uses-permission android:name=
"android.permission.SET_DEBUG_APP" />
<uses-permission android:name=
"android.permission.SET_ORIENTATION" />
<uses-permission android:name=
"android.permission.SET_POINTER_SPEED" />
<uses-permission android:name=
"android.permission.SET_PROCESS_LIMIT" />
<uses-permission android:name=
"android.permission.SET_TIME" />
<uses-permission android:name=
"android.permission.SET_TIME_ZONE" />
<uses-permission android:name=
"android.permission.SET_WALLPAPER" />
<uses-permission android:name=
"android.permission.SET_WALLPAPER_HINTS" />
<uses-permission android:name=
"android.permission.SIGNAL_PERSISTENT_PROCESSES" />
<uses-permission android:name=
"android.permission.STATUS_BAR" />
<uses-permission android:name=
"android.permission.SUBSCRIBED_FEEDS_READ" />
<uses-permission android:name=
```

```
"android.permission.SUBSCRIBED_FEEDS_WRITE" />
<uses-permission android:name=
"android.permission.SYSTEM_ALERT_WINDOW" />
<uses-permission android:name=
"android.permission.UPDATE_DEVICE_STATS" />
<uses-permission android:name=
"android.permission.USE_CREDENTIALS" />
<uses-permission android:name=
"android.permission.USE_SIP" />
<uses-permission android:name=
"android.permission.WAKE_LOCK" />
<uses-permission android:name=
"android.permission.WRITE_APN_SETTINGS" />
<uses-permission android:name=
"android.permission.WRITE_CALENDAR" />
<uses-permission android:name=
"android.permission.WRITE_CONTACTS" />
<uses-permission android:name=
"android.permission.WRITE_GSERVICES" />
<uses-permission android:name=
"android.permission.WRITE_HISTORY_BOOKMARKS" />
<uses-permission android:name=
"android.permission.WRITE_SECURE_SETTINGS" />
<uses-permission android:name=
"android.permission.WRITE_SETTINGS" />
<uses-permission android:name=
"android.permission.WRITE_SMS" />
<uses-permission android:name=
"android.permission.WRITE_SYNC_SETTINGS" />
<uses-permission android:name=
"android.permission.BIND_ACCESSIBILITY_SERVICE"/>
<uses-permission android:name=
"android.permission.BIND_TEXT_SERVICE"/>
<uses-permission android:name=
"android.permission.BIND_VPN_SERVICE"/>
<uses-permission android:name=
"android.permission.PERSISTENT_ACTIVITY"/>
<uses-permission android:name=
"android.permission.READ_CALL_LOG"/>
<uses-permission android:name=
"com.android.browser.permission.READ_HISTORY_BOOKMARKS"/>
<uses-permission android:name=
"android.permission.READ_PROFILE"/>
<uses-permission android:name=
```

```
"android.permission.READ_SOCIAL_STREAM"/>
<uses-permission android:name=
"android.permission.READ_USER_DICTIONARY"/>
<uses-permission android:name=
"com.android.alarm.permission.SET_ALARM"/>
<uses-permission android:name=
"android.permission.SET_PREFERRED_APPLICATIONS"/>
<uses-permission android:name=
"android.permission.WRITE_CALL_LOG"/>
<uses-permission android:name=
"com.android.browser.permission.WRITE_HISTORY_BOOKMARKS"/>
<uses-permission android:name=
"android.permission.WRITE_PROFILE"/>
<uses-permission android:name=
"android.permission.WRITE_SOCIAL_STREAM"/>
<uses-permission android:name=
"android.permission.WRITE_USER_DICTIONARY"/>
```

6.2 请求权限

在 Android SDK 版本 23 之后，权限需要在运行期内予以请求（但并非全部）。这意味着，需要从代码中请求权限。本节将描述如何从应用程序中执行这项任务。具体来说，当用户打开应用程序时，将请求获得 GPS 位置所需的权限。如果权限未获得批准，将向用户显示一个对话框以执行权限审批。

对此，打开 BaseActivity 类，并按照下列方式进行扩展：

```
abstract class BaseActivity : AppCompatActivity() {
  companion object {
  val REQUEST_GPS = 0
  ... }
  ...
  override fun onCreate(savedInstanceState: Bundle?) {
    super.onCreate(savedInstanceState)
    ...
    requestGpsPermissions() }
...
private fun requestGpsPermissions() {
  ActivityCompat.requestPermissions(
    this@BaseActivity,
    arrayOf(
      Manifest.permission.ACCESS_FINE_LOCATION,
```

```
     Manifest.permission.ACCESS_COARSE_LOCATION ),
     REQUEST_GPS ) }
     ...
override fun onRequestPermissionsResult(
  requestCode:
   Int, permissions: Array<String>, grantResults: IntArray ) {
     if (requestCode == REQUEST_GPS) {
      for (grantResult in grantResults)
      { if (grantResult == PackageManager.PERMISSION_GRANTED)
       { Log.i( tag, String.format( Locale.ENGLISH, "Permission
         granted [ %d ]", requestCode ) )
       }
       else {
         Log.e( tag, String.format( Locale.ENGLISH, "Permission
         not granted [ %d ]", requestCode ) )
       } } } } }
```

稍后将对上述代码进行逐行解释。

在 companion 对象中，我们定义了请求 ID，并等待该 ID 的结果。在 onCreate()方法中，调用了 requestGpsPermissions()方法，并在所定义的 ID 下执行授权请求。该权限请求的结果将用于 onRequestPermissionsResult()重载方法中。可以看到，这里采用了日志方式记录授权请求结果。当前，应用程序可检索 GPS 数据。

其他 Android 权限的处理过程也基本相同。构建并运行应用程序，随后将会显示如图 6.1 所示的授权画面。

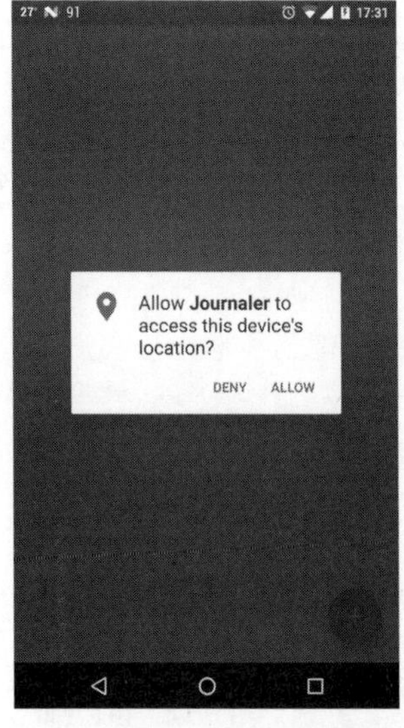

图 6.1

6.3 Kotlin 方案

如果应用程序请求了大量授权，且必须通过代码方式予以处理，情况又当如何？也就是说，我们有很多代码处理不同的权限请求，这意味着存在大量的样板代码！鉴于正在使用 Kotlin 这一工具，其将使得该处理过程变得更加简单。

创建名为 permission 的新数据包，随后生成两个新的 Kotlin 文件，即 PermissionCompatActivity 和 PermissionRequestCallback。

下列代码定义了授权请求回调：

```
package com.journaler.permission

interface PermissionRequestCallback {
  fun onPermissionGranted(permissions: List<String>)
  fun onPermissionDenied(permissions: List<String>)
}
```

这是在处理权限时所触发的回调。接下来定义权限 compat 活动，如下所示。

```
package com.journaler.permission

import android.content.pm.PackageManager
import android.support.v4.app.ActivityCompat
import android.support.v7.app.AppCompatActivity
import android.util.Log
import java.util.concurrent.ConcurrentHashMap
import java.util.concurrent.atomic.AtomicInteger

abstract class PermissionCompatActivity : AppCompatActivity() {

  private val tag = "Permissions extension"
  private val latestPermissionRequest = AtomicInteger()
  private val permissionRequests = ConcurrentHashMap<Int,
  List<String>>()
  private val permissionCallbacks =
   ConcurrentHashMap<List<String>, PermissionRequestCallback>()

  private val defaultPermissionCallback = object :
  PermissionRequestCallback {
    override fun onPermissionGranted(permissions: List<String>) {
```

```
            Log.i(tag, "Permission granted [ $permissions ]")
        }
        override fun onPermissionDenied(permissions: List<String>) {
            Log.e(tag, "Permission denied [ $permissions ]")
        }
    }

    fun requestPermissions(
        vararg permissions: String,
        callback: PermissionRequestCallback = defaultPermissionCallback
    ) {
        val id = latestPermissionRequest.incrementAndGet()
        val items = mutableListOf<String>()
        items.addAll(permissions)
        permissionRequests[id] = items
        permissionCallbacks[items] = callback
        ActivityCompat.requestPermissions(this, permissions, id)
    }

    override fun onRequestPermissionsResult(
        requestCode: Int,
        permissions: Array<String>,
        grantResults: IntArray
    ) {
        val items = permissionRequests[requestCode]
        items?.let {
            val callback = permissionCallbacks[items]
            callback?.let {
                var success = true
                for (x in 0..grantResults.lastIndex) {
                    val result = grantResults[x]
                    if (result != PackageManager.PERMISSION_GRANTED) {
                        success = false
                        break
                    }
                }
                if (success) {
                    callback.onPermissionGranted(items)
                } else {
                    callback.onPermissionDenied(items)
                }
            }
```

```
    }
  }
}
```

该类背后的理念是，将终端用户公开予 requestPermissions()方法，该方法接收可变的参数数量，表示当前所感兴趣的权限。我们可以传递可选的（optional）回调（刚刚定义的接口）。如果不传递自己的回调，代码将使用默认的回调机制。随后，在权限被处理后，将触发回调。只有在授予所有权限时，我们才认为权限处理成功。

下列代码更新 BaseActivity 类：

```
abstract class BaseActivity : PermissionCompatActivity() {
  ...
  override fun onCreate(savedInstanceState: Bundle?) {
     ...
     requestPermissions(
        Manifest.permission.ACCESS_FINE_LOCATION,
        Manifest.permission.ACCESS_COARSE_LOCATION
     )
  }
  ...
}
```

不难发现，这里从 BaseActivity 类中删除了之前所有与权限相关的代码，并通过 requestPermission()方法调用对其予以替换。

6.4 本章小结

本章内容稍显简短，但所涉及的内容却十分重要。权限对于每个 Android 应用程序来说均不可或缺，这对于用户和开发人员来说都是一种保护措施。根据具体需求，存在大量的彼此各异的授权方案。

第 7 章将继续探讨系统问题，读者将学习数据库的处理方法。

第 7 章　与数据库协同工作

在第6章讨论了访问Android系统特性时所需的重要的权限方面问题。在相关示例中，我们得到了地理位置权限。本章将介绍如何将数据插入数据库中，并插入从 Android 位置供应商那里所得到的位置数据。对此，需要定义适宜的数据库模式以及相应的管理类。除此之外，本章还将定义对应类以访问位置供应商，进而获得位置数据。

本章主要涉及以下主题：

- ❑ SQLite 简介。
- ❑ 描述数据库。
- ❑ CRUD 操作。

7.1　SQLite 简介

当持久化应用程序数据时，将需要使用数据库。在 Android 中，对于离线存储，可以采用 SQLite。

SQLite 具有“开箱即用”这一特性，这意味着，SQLite 包含于 Android 框架中。

SQLite 的优点是功能强大、速度快且非常可靠。如果用户发现任何问题，一般很容易找到解决方案，因为社区中很可能已经有人解决了这些问题。SQLite 是一个自包含的、嵌入式的、功能齐全的公共域 SQL 数据库引擎。

本章将采用 SQLite 存储所有的 Todo 和 Note。对此，可定义相应的数据库、访问机制和数据管理方案。当然，我们不会直接公开数据库实例，而是对其进行适当的封装，以便插入、更新、查询或删除数据。

7.2　描述数据库

首先，本节将定义包含适宜数据类型的表和列以描述数据库。除此之外，还将定义所表达数据的简单模型。因此，下面将创建一个名为 database 的新数据库，如下所示。

```
com.journaler.database
```

接下来，定义一个名为 DbModel 的类。DbModel 类表示为全部应用程序数据库模型的矩阵，且仅包含 ID——ID 是一个强制型字段，并可用作主键。这里，应确保 DbModel 类按照下列方式加以定义：

```
package com.journaler.database

abstract class DbModel {
 abstract var id: Long
}
```

当定义起始点时，将定义真正包含数据的数据类。在现有的名为 model 数据包中，创建新的 DbEntry 类、Note 类和 Todo 类，其中，Note 类和 Todo 类扩展了 Entry 类；而 Entry 类则扩展了 DbModel 类。

相应地，Entry 类定义如下：

```
package com.journaler.model

import android.location.Location
import com.journaler.database.DbModel

abstract class Entry(
 var title: String,
 var message: String,
 var location: Location
) : DbModel()
Note class:
package com.journaler.model

import android.location.Location

class Note(
 title: String,
 message: String,
 location: Location
) : Entry(
 title,
 message,
 location
 ) {
   override var id = 0L
 }
```

可以看到，我们将当前地理位置连同 title 和 message 内容一起作为存储在 Note 中的信息。此外，这里还重载了 ID。鉴于新实例化的 note 尚未存储于数据库中，因而其对应 ID 为 0。待存储完毕后，将更新为从数据库中获取的 ID 值。

Todo 类定义如下：

```
package com.journaler.model

import android.location.Location

class Todo(
  title: String,
  message: String,
  location: Location,
  var scheduledFor: Long
) : Entry(
  title,
  message,
  location
) {
  override var id = 0L
}
```

与 Note 类相比，Todo 类包含了一个附加字段，即 todo 被调度时的时间戳。

在定义了数据模型后，接下来将描述数据库。对此，需要定义针对数据库实例化的数据库辅助类。另外，该数据库辅助类需要扩展 Android 的 SQLiteOpenHelper 类。下面定义一个 DbHelper 类，并确保该类扩展 SQLiteOpenHelper 类，如下所示。

```
package com.journaler.database

import android.database.sqlite.SQLiteDatabase
import android.database.sqlite.SQLiteOpenHelper
import android.util.Log
import com.journaler.Journaler

class DbHelper(val dbName: String, val version: Int) :
SQLiteOpenHelper(
  Journaler.ctx, dbName, null, version
) {

  companion object {
    val ID: String = "_id"
    val TABLE_TODOS = "todos"
```

```
  val TABLE_NOTES = "notes"
  val COLUMN_TITLE: String = "title"
  val COLUMN_MESSAGE: String = "message"
  val COLUMN_SCHEDULED: String = "scheduled"
  val COLUMN_LOCATION_LATITUDE: String = "latitude"
  val COLUMN_LOCATION_LONGITUDE: String = "longitude"
}

private val tag = "DbHelper"

private val createTableNotes = """
  CREATE TABLE if not exists $TABLE_NOTES
    (
      $ID integer PRIMARY KEY autoincrement,
      $COLUMN_TITLE text,
      $COLUMN_MESSAGE text,
      $COLUMN_LOCATION_LATITUDE real,
      $COLUMN_LOCATION_LONGITUDE real
    )
    """

private val createTableTodos = """
  CREATE TABLE if not exists $TABLE_TODOS
    (
      $ID integer PRIMARY KEY autoincrement,
      $COLUMN_TITLE text,
      $COLUMN_MESSAGE text,
      $COLUMN_SCHEDULED integer,
      $COLUMN_LOCATION_LATITUDE real,
      $COLUMN_LOCATION_LONGITUDE real
    )
    """

override fun onCreate(db: SQLiteDatabase) {
  Log.d(tag, "Database [ CREATING ]")
  db.execSQL(createTableNotes)
  db.execSQL(createTableTodos)
  Log.d(tag, "Database [ CREATED ]")
}

override fun onUpgrade(db: SQLiteDatabase?, oldVersion: Int,
newVersion: Int) {
```

```
    // Ignore for now.
  }

}
```

SQLiteOpenHelper 类的 companion 对象包含了针对表和列名的定义。此外，还针对表生成操作定义了 SQL。最后，SQL 将在 onCreate()方法中被执行。7.3 节将进一步讨论数据库管理以及数据插入操作。

7.3　CRUD 操作

CURD 操作涵盖了数据的创建、插入更新、删除、选择操作，并通过一个名为 Crud 的接口加以定义，同时兼具通用性。对此，在 database 数据库中创建一个新接口，并确保包含所有的 CRUD 操作，如下所示。

```
interface Crud<T> where T : DbModel {

  companion object {
    val BROADCAST_ACTION = "com.journaler.broadcast.crud"
    val BROADCAST_EXTRAS_KEY_CRUD_OPERATION_RESULT = "crud_result"
  }

  /**
  * Returns the ID of inserted item.
  */
  fun insert(what: T): Long

  /**
  * Returns the list of inserted IDs.
  */
  fun insert(what: Collection<T>): List<Long>

  /**
  * Returns the number of updated items.
  */
  fun update(what: T): Int

  /**
  * Returns the number of updated items.
  */
```

```
    fun update(what: Collection<T>): Int

    /**
    * Returns the number of deleted items.
    */
    fun delete(what: T): Int

    /**
    * Returns the number of deleted items.
    */
    fun delete(what: Collection<T>): Int

    /**
    * Returns the list of items.
    */
    fun select(args: Pair<String, String>): List<T>

    /**
    * Returns the list of items.
    */
    fun select(args: Collection<Pair<String, String>>): List<T>

    /**
    * Returns the list of items.
    */
    fun selectAll(): List<T>

}
```

当执行 CRUD 操作时，存在两个方法版本。其中，第一个版本接收实例集合；第二个版本则接收单一项。下面通过定义一个名为 Db 的 Kotlin 对象实现 CRUD 的具体化操作。也就是说，创建一个对象，以使具体化操作形成一个完美的单例。这里，Db 对象须实现 Crud 接口，如下所示。

```
package com.journaler.database

import android.content.ContentValues
import android.location.Location
import android.util.Log
import com.journaler.model.Note
import com.journaler.model.Todo
```

```
object Db {

  private val tag = "Db"
  private val version = 1
  private val name = "students"

  val NOTE = object : Crud<Note> {
    // Crud implementations
  }

  val TODO = object : Crud<NoteTodo {
    // Crud implementations
  }
}
```

7.3.1　插入操作

insert 操作中将向数据库中添加新数据，其实现过程如下所示。

```
val NOTE = object : Crud<Note> {
  ...
  override fun insert(what: Note): Long {
    val inserted = insert(listOf(what))
    if (!inserted.isEmpty()) return inserted[0]
    return 0
  }

  override fun insert(what: Collection<Note>): List<Long> {
    val db = DbHelper(name, version).writableDatabase
    db.beginTransaction()
    var inserted = 0
    val items = mutableListOf<Long>()
    what.forEach { item ->
      val values = ContentValues()
      val table = DbHelper.TABLE_NOTES
      values.put(DbHelper.COLUMN_TITLE, item.title)
      values.put(DbHelper.COLUMN_MESSAGE, item.message)
      values.put(DbHelper.COLUMN_LOCATION_LATITUDE,
        item.location.latitude)
      values.put(DbHelper.COLUMN_LOCATION_LONGITUDE,
        item.location.longitude)
      val id = db.insert(table, null, values)
```

```
            if (id > 0) {
              items.add(id)
              Log.v(tag, "Entry ID assigned [ $id ]")
              inserted++
            }
          }
          val success = inserted == what.size
          if (success) {
              db.setTransactionSuccessful()
          } else {
              items.clear()
          }
          db.endTransaction()
          db.close()
          return items
        }
        ...
}
...
val TODO = object : Crud<Todo> {
  ...
  override fun insert(what: Todo): Long {
    val inserted = insert(listOf(what))
    if (!inserted.isEmpty()) return inserted[0]
    return 0
  }

  override fun insert(what: Collection<Todo>): List<Long> {
    val db = DbHelper(name, version).writableDatabase
    db.beginTransaction()
    var inserted = 0
    val items = mutableListOf<Long>()
    what.forEach { item ->
      val table = DbHelper.TABLE_TODOS
      val values = ContentValues()
      values.put(DbHelper.COLUMN_TITLE, item.title)
      values.put(DbHelper.COLUMN_MESSAGE, item.message)
      values.put(DbHelper.COLUMN_LOCATION_LATITUDE,
        item.location.latitude)
      values.put(DbHelper.COLUMN_LOCATION_LONGITUDE,
        item.location.longitude)
      values.put(DbHelper.COLUMN_SCHEDULED, item.scheduledFor)
```

```
    val id = db.insert(table, null, values)
      if (id > 0) {
        item.id = id
        Log.v(tag, "Entry ID assigned [ $id ]")
        inserted++
      }
    }
    val success = inserted == what.size
    if (success) {
        db.setTransactionSuccessful()
    } else {
        items.clear()
    }
    db.endTransaction()
    db.close()
    return items
  }
  ...
}
...
```

7.3.2　更新操作

update 操作将更新数据库中现有的数据，其实现过程如下所示。

```
val NOTE = object : Crud<Note> {
  ...
  override fun update(what: Note) = update(listOf(what))

  override fun update(what: Collection<Note>): Int {
    val db = DbHelper(name, version).writableDatabase
    db.beginTransaction()
    var updated = 0
    what.forEach { item ->
      val values = ContentValues()
      val table = DbHelper.TABLE_NOTES
      values.put(DbHelper.COLUMN_TITLE, item.title)
      values.put(DbHelper.COLUMN_MESSAGE, item.message)
      values.put(DbHelper.COLUMN_LOCATION_LATITUDE,
      item.location.latitude)
      values.put(DbHelper.COLUMN_LOCATION_LONGITUDE,
      item.location.longitude)
```

```
    db.update(table, values, "_id = ?",
    arrayOf(item.id.toString()))
      updated++
   }
   val result = updated == what.size
   if (result) {
      db.setTransactionSuccessful()
   } else {
      updated = 0
   }
   db.endTransaction()
   db.close()
   return updated
  }
  ...
}
...
val TODO = object : Crud<Todo> {
  ...
  override fun update(what: Todo) = update(listOf(what))
  override fun update(what: Collection<Todo>): Int {
    val db = DbHelper(name, version).writableDatabase
    db.beginTransaction()
    var updated = 0
    what.forEach { item ->
      val table = DbHelper.TABLE_TODOS
      val values = ContentValues()
      values.put(DbHelper.COLUMN_TITLE, item.title)
      values.put(DbHelper.COLUMN_MESSAGE, item.message)
      values.put(DbHelper.COLUMN_LOCATION_LATITUDE,
      item.location.latitude)
      values.put(DbHelper.COLUMN_LOCATION_LONGITUDE,
      item.location.longitude)
      values.put(DbHelper.COLUMN_SCHEDULED, item.scheduledFor)
      db.update(table, values, "_id = ?",
      arrayOf(item.id.toString()))
        updated++
    }
    val result = updated == what.size
    if (result) {
       db.setTransactionSuccessful()
    } else {
```

```
        updated = 0
      }
      db.endTransaction()
      db.close()
      return updated
    }
    ...
  }
...
```

7.3.3 删除操作

delete 操作将从数据库中移除现有的数据，其实现过程如下所示。

```
val NOTE = object : Crud<Note> {
  ...
  override fun delete(what: Note): Int = delete(listOf(what))
  override fun delete(what: Collection<Note>): Int {
    val db = DbHelper(name, version).writableDatabase
    db.beginTransaction()
    val ids = StringBuilder()
    what.forEachIndexed { index, item ->
      ids.append(item.id.toString())
      if (index < what.size - 1) {
        ids.append(", ")
      }
    }
    val table = DbHelper.TABLE_NOTES
    val statement = db.compileStatement(
        "DELETE FROM $table WHERE ${DbHelper.ID} IN ($ids);"
    )
    val count = statement.executeUpdateDelete()
    val success = count > 0
    if (success) {
      db.setTransactionSuccessful()
      Log.i(tag, "Delete [ SUCCESS ][ $count ][ $statement ]")
    } else {
        Log.w(tag, "Delete [ FAILED ][ $statement ]")
    }
    db.endTransaction()
    db.close()
    return count
```

```
  }
  ...
}
...
val TODO = object : Crud<Todo> {
  ...
  override fun delete(what: Todo): Int = delete(listOf(what))
  override fun delete(what: Collection<Todo>): Int {
    val db = DbHelper(name, version).writableDatabase
    db.beginTransaction()
    val ids = StringBuilder()
    what.forEachIndexed { index, item ->
      ids.append(item.id.toString())
      if (index < what.size - 1) {
        ids.append(", ")
      }
    }
    val table = DbHelper.TABLE_TODOS
    val statement = db.compileStatement(
        "DELETE FROM $table WHERE ${DbHelper.ID} IN ($ids);"
    )
    val count = statement.executeUpdateDelete()
    val success = count > 0
    if (success) {
      db.setTransactionSuccessful()
      Log.i(tag, "Delete [ SUCCESS ][ $count ][ $statement ]")
    } else {
        Log.w(tag, "Delete [ FAILED ][ $statement ]")
    }
    db.endTransaction()
    db.close()
    return count
  }
  ...
}
...
```

7.3.4 选择操作

select 操作将从数据库中读取并返回数据，其实现过程如下所示。

```
val NOTE = object : Crud<Note> {
```

```
...
override fun select(
    args: Pair<String, String>
): List<Note> = select(listOf(args))

override fun select(args: Collection<Pair<String, String>>):
List<Note> {
  val db = DbHelper(name, version).writableDatabase
  val selection = StringBuilder()
  val selectionArgs = mutableListOf<String>()
  args.forEach { arg ->
    selection.append("${arg.first} == ?")
    selectionArgs.add(arg.second)
  }
  val result = mutableListOf<Note>()
  val cursor = db.query(
    true,
    DbHelper.TABLE_NOTES,
    null,
    selection.toString(),
    selectionArgs.toTypedArray(),
    null, null, null, null
  )
  while (cursor.moveToNext()) {
    val id = cursor.getLong(cursor.getColumnIndexOrThrow
    (DbHelper.ID))
    val titleIdx = cursor.getColumnIndexOrThrow
    (DbHelper.COLUMN_TITLE)
    val title = cursor.getString(titleIdx)
    val messageIdx = cursor.getColumnIndexOrThrow
    (DbHelper.COLUMN_MESSAGE)
    val message = cursor.getString(messageIdx)
    val latitudeIdx = cursor.getColumnIndexOrThrow(
        DbHelper.COLUMN_LOCATION_LATITUDE
    )
    val latitude = cursor.getDouble(latitudeIdx)
    val longitudeIdx = cursor.getColumnIndexOrThrow(
      DbHelper.COLUMN_LOCATION_LONGITUDE
    )
    val longitude = cursor.getDouble(longitudeIdx)
    val location = Location("")
    location.latitude = latitude
```

```
    location.longitude = longitude
    val note = Note(title, message, location)
    note.id = id
    result.add(note)
  }
  cursor.close()
  return result
}

override fun selectAll(): List<Note> {
  val db = DbHelper(name, version).writableDatabase
  val result = mutableListOf<Note>()
  val cursor = db.query(
    true,
    DbHelper.TABLE_NOTES,
    null, null, null, null, null, null, null
  )
  while (cursor.moveToNext()) {
    val id = cursor.getLong(cursor.getColumnIndexOrThrow
    (DbHelper.ID))
    val titleIdx = cursor.getColumnIndexOrThrow
    (DbHelper.COLUMN_TITLE)
    val title = cursor.getString(titleIdx)
    val messageIdx = cursor.getColumnIndexOrThrow
    (DbHelper.COLUMN_MESSAGE)
    val message = cursor.getString(messageIdx)
    val latitudeIdx = cursor.getColumnIndexOrThrow(
      DbHelper.COLUMN_LOCATION_LATITUDE
    )
    val latitude = cursor.getDouble(latitudeIdx)
    val longitudeIdx = cursor.getColumnIndexOrThrow(
      DbHelper.COLUMN_LOCATION_LONGITUDE
    )
    val longitude = cursor.getDouble(longitudeIdx)
    val location = Location("")
    location.latitude = latitude
    location.longitude = longitude
    val note = Note(title, message, location)
    note.id = id
    result.add(note)
  }
  cursor.close()
```

```
    return result
    }
    ...
}
...
val TODO = object : Crud<Todo> {
  ...
  override fun select(args: Pair<String, String>): List<Todo> =
  select(listOf(args))

  override fun select(args: Collection<Pair<String, String>>):
  List<Todo> {
    val db = DbHelper(name, version).writableDatabase
    val selection = StringBuilder()
    val selectionArgs = mutableListOf<String>()
    args.forEach { arg ->
      selection.append("${arg.first} == ?")
      selectionArgs.add(arg.second)
    }
    val result = mutableListOf<Todo>()
    val cursor = db.query(
        true,
        DbHelper.TABLE_NOTES,
        null,
        selection.toString(),
        selectionArgs.toTypedArray(),
        null, null, null, null
    )
    while (cursor.moveToNext()) {
      val id = cursor.getLong(cursor.getColumnIndexOrThrow
      (DbHelper.ID))
      val titleIdx = cursor.getColumnIndexOrThrow
      (DbHelper.COLUMN_TITLE)
      val title = cursor.getString(titleIdx)
      val messageIdx = cursor.getColumnIndexOrThrow
      (DbHelper.COLUMN_MESSAGE)
      val message = cursor.getString(messageIdx)
      val latitudeIdx = cursor.getColumnIndexOrThrow(
          DbHelper.COLUMN_LOCATION_LATITUDE
      )
      val latitude = cursor.getDouble(latitudeIdx)
      val longitudeIdx = cursor.getColumnIndexOrThrow(
```

```
        DbHelper.COLUMN_LOCATION_LONGITUDE
      )
      val longitude = cursor.getDouble(longitudeIdx)
      val location = Location("")
      val scheduledForIdx = cursor.getColumnIndexOrThrow(
         DbHelper.COLUMN_SCHEDULED
      )
      val scheduledFor = cursor.getLong(scheduledForIdx)
      location.latitude = latitude
      location.longitude = longitude
      val todo = Todo(title, message, location, scheduledFor)
      todo.id = id
      result.add(todo)
    }
    cursor.close()
    return result
  }
  override fun selectAll(): List<Todo> {
    val db = DbHelper(name, version).writableDatabase
    val result = mutableListOf<Todo>()
    val cursor = db.query(
       true,
       DbHelper.TABLE_NOTES,
       null, null, null, null, null, null, null
    )
    while (cursor.moveToNext()) {
      val id = cursor.getLong(cursor.getColumnIndexOrThrow
      (DbHelper.ID))
      val titleIdx = cursor.getColumnIndexOrThrow
      (DbHelper.COLUMN_TITLE)
      val title = cursor.getString(titleIdx)
      val messageIdx = cursor.getColumnIndexOrThrow
      (DbHelper.COLUMN_MESSAGE)
      val message = cursor.getString(messageIdx)
      val latitudeIdx = cursor.getColumnIndexOrThrow(
         DbHelper.COLUMN_LOCATION_LATITUDE
      )
      val latitude = cursor.getDouble(latitudeIdx)
      val longitudeIdx = cursor.getColumnIndexOrThrow(
         DbHelper.COLUMN_LOCATION_LONGITUDE
      )
      val longitude = cursor.getDouble(longitudeIdx)
```

```
      val location = Location("")
      val scheduledForIdx = cursor.getColumnIndexOrThrow(
          DbHelper.COLUMN_SCHEDULED
      )
      val scheduledFor = cursor.getLong(scheduledForIdx)
      location.latitude = latitude
      location.longitude = longitude
      val todo = Todo(title, message, location, scheduledFor)
      todo.id = id
      result.add(todo)
    }
    cursor.close()
    return result
  }
  ...
}
...
```

通过使用 DbHelper 类，每项 CRUD 操作均可获得一个数据库实例，并通过 CRUD 机制对其加以使用，而非直接将其公开。在每次操作后，数据库将被关闭。这里，我们仅可得到一个可读的数据库；在当前示例中，通过访问 writableDatabase 可得到一个 writableDatabase 实例。这意味着，此处将通过调用数据库实例上的 beginTransaction()方法予以启动。相应地，调用 endTransaction()方法则可完成事务——如果未在之前调用 setTransactionSuccessful()方法，则数据库不会发生任何变化。如前所述，每项 CRUD 操作存在两个版本。其中，第一个版本包含了主实现；第二个版本则仅传递实例。当执行数据库的插入操作时，需要注意的是，这里将使用数据库实例上的 insert()方法，该方法接收所插入的表名，以及表示数据的内容值（ContentValues 类）。update 和 delete 操作则较为类似。对此，我们可使用 update()方法和 delete()方法。在当前示例中，针对数据移除行为，我们采用了包含删除 SQL 查询的 compileStatement()方法。

注意：

这里所提供的代码稍显复杂，读者仅需将注意力集中于数据库部分。同时，我们也鼓励读者在前述 Android 数据库类的基础上创建自己版本的数据库管理类。

7.3.5　整合方案

本节将通过数据库类执行 CRUD 操作，并扩展应用程序以创建 Note，其间的主要工作是执行插入操作。

在向数据库中插入数据之前，首先需要提供一种机制获取当前用户的地理位置，这对于 Note 和 Todo 来说均不可或缺。对此，创建一个名为 LocationProvider 的新对象，并将其置于 location 数据包中，如下所示。

```
object LocationProvider {
  private val tag = "Location provider"
  private val listeners = CopyOnWriteArrayList
  <WeakReference<LocationListener>>()

  private val locationListener = object : LocationListener {
    ...
  }

  fun subscribe(subscriber: LocationListener): Boolean {
    val result = doSubscribe(subscriber)
    turnOnLocationListening()
    return result
  }

  fun unsubscribe(subscriber: LocationListener): Boolean {
    val result = doUnsubscribe(subscriber)
    if (listeners.isEmpty()) {
        turnOffLocationListening()
    }
    return result
  }

  private fun turnOnLocationListening() {
    ...
  }

  private fun turnOffLocationListening() {
    ...
  }

  private fun doSubscribe(listener: LocationListener): Boolean {
    ...
  }

  private fun doUnsubscribe(listener: LocationListener): Boolean {
    ...
  }
}
```

此处公开了 LocationProvider 对象的主结构，接下来查看该实现的其余部分。其中，locationListener 实例代码如下所示。

```
private val locationListener = object : LocationListener {
    override fun onLocationChanged(location: Location) {
        Log.i(
                tag,
                String.format(
                        Locale.ENGLISH,
                        "Location [ lat: %s ][ long: %s ]",
                        location.latitude, location.longitude
                )
        )
        val iterator = listeners.iterator()
        while (iterator.hasNext()) {
            val reference = iterator.next()
            val listener = reference.get()
            listener?.onLocationChanged(location)
        }
    }

    override fun onStatusChanged(provider: String, status: Int,
    extras: Bundle) {
        Log.d(
                tag,
                String.format(Locale.ENGLISH, "Status changed [ %s
                ][ %d ]", provider, status)
        )
        val iterator = listeners.iterator()
        while (iterator.hasNext()) {
            val reference = iterator.next()
            val listener = reference.get()
            listener?.onStatusChanged(provider, status, extras)
        }
    }

    override fun onProviderEnabled(provider: String) {
        Log.i(tag, String.format("Provider [ %s ][ ENABLED ]",provider))
        val iterator = listeners.iterator()
        while (iterator.hasNext()) {
            val reference = iterator.next()
            val listener = reference.get()
```

```
            listener?.onProviderEnabled(provider)
        }
    }

    override fun onProviderDisabled(provider: String) {
        Log.i(tag, String.format("Provider [ %s ][ ENABLED ]",provider))
        val iterator = listeners.iterator()
        while (iterator.hasNext()) {
            val reference = iterator.next()
            val listener = reference.get()
            listener?.onProviderDisabled(provider)
        }
    }
}
```

LocationListener 表示为 Android 的接口，其功能将在 location 事件中被扩展，基本上，其具体操作会将所有相关事件通知到所有的订阅方。这里，较为重要的部分是 onLocationChanged()方法，如下所示。

```
turnOnLocationListening():

private fun turnOnLocationListening() {
   Log.v(tag, "We are about to turn on location listening.")
   val ctx = Journaler.ctx
   if (ctx != null) {
         Log.v(tag, "We are about to check location permissions.")

         val permissionsOk =
         ActivityCompat.checkSelfPermission(ctx,
         Manifest.permission.ACCESS_FINE_LOCATION) ==
         PackageManager.PERMISSION_GRANTED
         &&
         ActivityCompat.checkSelfPermission(ctx,
         Manifest.permission.ACCESS_COARSE_LOCATION) ==
         PackageManager.PERMISSION_GRANTED

         if (!permissionsOk) {
              throw IllegalStateException(
              "Permissions required [ ACCESS_FINE_LOCATION ]
                [ ACCESS_COARSE_LOCATION ]"
              )
         }
         Log.v(tag, "Location permissions are ok.
```

```
    We are about to request location changes.")
    val locationManager =
    ctx.getSystemService(Context.LOCATION_SERVICE)
    as LocationManager

    val criteria = Criteria()
    criteria.accuracy = Criteria.ACCURACY_FINE
    criteria.powerRequirement = Criteria.POWER_HIGH
    criteria.isAltitudeRequired = false
    criteria.isBearingRequired = false
    criteria.isSpeedRequired = false
    criteria.isCostAllowed = true

    locationManager.requestLocationUpdates(
            1000, 1F, criteria, locationListener,
            Looper.getMainLooper()
    )
    } else {
      Log.e(tag, "No application context available.")
  }
}
```

当打开地理位置监听时，需要检测是否具备相关权限。若已授权，则可获取 Android 的 LocationManager，并针对地理位置更新而定义了 Criteria。最后，可通过传递以下参数来请求位置更新：

- long minTime。
- float minDistance。
- Criteria criteria。
- LocationListener listener。
- Looper looper。

可以看到，此处传递了 LocationListener，进而将 location 事件通知至所有订阅的第三方，如下所示。

```
turnOffLocationListening():private fun turnOffLocationListening()
{
  Log.v(tag, "We are about to turn off location listening.")
  val ctx = Journaler.ctx
  if (ctx != null) {
    val locationManager =
    ctx.getSystemService(Context.LOCATION_SERVICE)
```

```
    as LocationManager

    locationManager.removeUpdates(locationListener)
  } else {
      Log.e(tag, "No application context available.")
  }
}
```

简单地移除监听器 instance.doSubscribe()，即可终止地理位置的监听行为，如下所示。

```
private fun doSubscribe(listener: LocationListener): Boolean {
  val iterator = listeners.iterator()
  while (iterator.hasNext()) {
    val reference = iterator.next()
    val refListener = reference.get()
    if (refListener != null && refListener === listener) {
          Log.v(tag, "Already subscribed: " + listener)
          return false
     }
  }
   listeners.add(WeakReference(listener))
   Log.v(tag, "Subscribed, subscribers count: " + listeners.size)
   return true
}
```

doUnsubscribe()方法代码如下所示。

```
private fun doUnsubscribe(listener: LocationListener): Boolean {
  var result = true
  val iterator = listeners.iterator()
  while (iterator.hasNext()) {
     val reference = iterator.next()
     val refListener = reference.get()
     if (refListener != null && refListener === listener) {
         val success = listeners.remove(reference)
         if (!success) {
             Log.w(tag, "Couldn't un subscribe, subscribers
             count: " + listeners.size)
         } else {
             Log.v(tag, "Un subscribed, subscribers count: " +
             listeners.size)
         }
         if (result) {
             result = success
```

```
            }
        }
    }
    return result
}
```

针对第三方，上述两个方法分别负责订阅和解除订阅位置更新。

接下来打开 NoteActivity 类，并按照下列方式对其进行扩展：

```
class NoteActivity : ItemActivity() {
  private var note: Note? = null
  override val tag = "Note activity"
  private var location: Location? = null
  override fun getLayout() = R.layout.activity_note
  private val textWatcher = object : TextWatcher {
    override fun afterTextChanged(p0: Editable?) {
      updateNote()
    }

    override fun beforeTextChanged(p0: CharSequence?, p1: Int, p2:
    Int, p3: Int) {}
    override fun onTextChanged(p0: CharSequence?, p1: Int, p2:
    Int, p3: Int) {}
  }

  private val locationListener = object : LocationListener {
    override fun onLocationChanged(p0: Location?) {
      p0?.let {
        LocationProvider.unsubscribe(this)
        location = p0
        val title = getNoteTitle()
        val content = getNoteContent()
        note = Note(title, content, p0)
        val task = object : AsyncTask<Note, Void, Boolean>() {
          override fun doInBackground(vararg params: Note?):
          Boolean {
            if (!params.isEmpty()) {
              val param = params[0]
              param?.let {
                return Db.NOTE.insert(param) > 0
              }
            }
            return false
```

```
        }

        override fun onPostExecute(result: Boolean?) {
          result?.let {
            if (result) {
               Log.i(tag, "Note inserted.")
            } else {
               Log.e(tag, "Note not inserted.")
            }
          }
        }
      }
      task.execute(note)
    }
  }

  override fun onStatusChanged(p0: String?, p1: Int, p2:
  Bundle?) {}
  override fun onProviderEnabled(p0: String?) {}
  override fun onProviderDisabled(p0: String?) {}
}

override fun onCreate(savedInstanceState: Bundle?) {
  super.onCreate(savedInstanceState)
  note_title.addTextChangedListener(textWatcher)
  note_content.addTextChangedListener(textWatcher)
}

private fun updateNote() {
  if (note == null) {
    if (!TextUtils.isEmpty(getNoteTitle()) &&
    !TextUtils.isEmpty(getNoteContent())) {
      LocationProvider.subscribe(locationListener)
    }
  } else {
    note?.title = getNoteTitle()
    note?.message = getNoteContent()
    val task = object : AsyncTask<Note, Void, Boolean>() {
      override fun doInBackground(vararg params: Note?):
      Boolean {
        if (!params.isEmpty()) {
          val param = params[0]
```

```
          param?.let {
            return Db.NOTE.update(param) > 0
          }
        }
        return false
      }

      override fun onPostExecute(result: Boolean?) {
        result?.let {
          if (result) {
            Log.i(tag, "Note updated.")
          } else {
            Log.e(tag, "Note not updated.")
          }
        }
      }
    }
    task.execute(note)
  }
}

private fun getNoteContent(): String {
  return note_content.text.toString()
}

private fun getNoteTitle(): String {
  return note_title.text.toString()
}
}
```

此处添加了两个字段。其中，第一个字段包含了当前正在编辑的 Note 实例；第二个字段则加载与当前用户地理位置相关的信息，并于随后定义了一个 TextWatcher 实例。TextWatcher 表示分配与 EditText 视图的一个监听器。每次产生变化时，相应的更新方法即会被触发。该方法将创建一个新的 Note 类，并将其持久化至数据库中（如果不存在），或者执行数据更新操作（如果存在）。

鉴于在没有位置数据可用之前，我们并不会插入 Note，因而这里定义了 locationListener，将接收到的地理位置置入 location 字段并取消订阅。随后将获取当前的 Note 标题值及其主要内容，进而生成一个新的 Note 实例。考虑到数据库操作可能会占用些许时间，因而可采用异步方式对其予以执行。对此，可使用 AsyncTask 类。AsyncTask 类是一个 Android 类，旨在执行大多数异步操作。该类定义了输入类型、进程类型和结果类型。在当前示例中，输入类型为 Note，且当前并不需要使用进程类型，但结果类型须为 Boolean，表示操作是否成功。

doInBackground()方法负责完成主要工作，对应结果则在 onPostExecute()方法中进行处理。可以看到，我们利用了刚刚定义的数据库管理类，并于后台执行了插入操作。

下一步是将 textWatcher 分配与 onCreate()方法中的 EditText 视图，并于随后定义最为重要的方法，即 updateNote()方法。该方法将更新现有的 Note，或者插入新的 Note（如果不存在）。再次，使用 AsyncTask 类于后台执行某项操作。

构建、运行应用程序，尝试插入 Note 并查看 Logcat。其中，与数据库相关的日志内容如下所示。

```
I/Note activity: Note inserted.
I/Note activity: Note updated.
I/Note activity: Note updated.
I/Note activity: Note updated.
```

至此，我们在 Android 中成功地实现了第一个数据库。这里，建议读者进一步对 CRUD 操作代码进行扩展，以确保 NoteActivity 支持 select 和 delete 操作。

7.4 本 章 小 结

本章阐述了如何在 Android 中实现复杂数据的持久化操作。数据库是应用程序的核心内容，Journaler 程序也不例外。本章讨论了 SQLite 数据库上所执行的全部 CRUD 操作，并逐一对其进行实现。在第 8 章中，还将对相对简单的数据介绍另一种持久化机制，即共享偏好设置，并以此保存简单的小型数据。

第 8 章　Android 偏好设置

第 7 章讲解了如何处理存储于 SQLite 数据库中的复杂数据，而本章则考查一些相对简单的数据形式，并通过特定的示例展示 Android 共享偏好设置的使用方式。

假设需要记住 ViewPager 类最后一个页面的位置，且每次应用程序启动时打开该页面。对此，可采用共享偏好设置记住该页面并持久化这一信息，进而在必要时对其进行检索。

本章主要涉及以下主题：

- Android 偏好设置的含义及其使用方式。
- 定义自己的偏好设置管理器。

8.1　Android 偏好设置的含义

应用程序的偏好设置称作共享偏好设置，并通过 Android 的各项偏好设置机制进行持久化和检索。各项偏好设置自身表示为 XML 数据，并可通过 Android 及其 API 予以访问和修改。关于偏好设置的检索和保存，Android 负责处理全部工作。另外，Android 还对此类偏好设置提供了相关机制，以实现私有、隐藏和公共访问。对于偏好设置管理，Android SDK 涵盖了较大的类集合。另外，一些抽象机制使得用户不必局限于默认的 XML，用户还可创建自己的持久化层。

8.2　使用方式

当使用共享偏好设置时，需要从当前上下文中获取 SharedPreferences 实例，如下所示。

```
val prefs = ctx.getSharedPreferences(key, mode)
```

这里，key 表示为一个 String 参数，并对共享偏好设置实例进行命名。系统中的 XML 文件也将包含这一名称。这表示为 Context 类中的有效模式（操作模式），如下所示。

- MODE_PRIVATE：表示为默认模式，生成后的文件仅可通过调用应用程序予以访问。

- MODE_WORLD_READABLE：表示为已弃用。
- MODE_WORLD_WRITEABLE：表示为已弃用。

随后，可通过下列方式对其进行存储和检索：

```
val value = prefs.getString("key", "default value")
```

针对全部创建的数据类型，还存在类似的 getter 方法。

8.2.1　编辑（存储）偏好设置

下面首先查看一个偏好设置编辑示例，如下所示。

```
preferences.edit().putString("key", "balue").commit()
```

注意：

commit()方法将即刻执行操作，而 apply()方法则在后台运行。当使用 commit()方法时，不要获取、操控应用程序主线程中的共享偏好设置。另外，应确保全部写入和读取操作在后台被执行。对此，可采用 AsyncTask 类；或者采用 apply()方法，而非 commit()方法。

8.2.2　移除偏好设置

当移除偏好设置时，可使用 remove()方法，如下所示。

```
prefs.edit().remove("key").commit()
```

注意，不要通过空数据覆写方式移除偏好设置。例如，利用 null 覆写整数；或者利用空字符串覆写字符串。

8.3　定义自己的设置管理器

本节将构建相应的机制以获取共享偏好设置。

对此，创建一个名为 preferences 的新数据包，并将全部关联代码置于该包内。对于共享偏好设置管理，需要定义以下 3 个类：

- PreferencesProviderAbstract：该类表示为基本的抽象，进而提供了 SharedPreferences 的访问行为。
- PreferencesConfiguration：该类负责描述尝试实例化的偏好设置。
- PreferencesProvider：该类表示为 PreferencesProviderAbstract 实现。

下面分别对每个类加以定义。

PreferencesProviderAbstract 代码如下所示。

```
package com.journaler.perferences

import android.content.Context
import android.content.SharedPreferences

abstract class PreferencesProviderAbstract {
  abstract fun obtain(configuration: PreferencesConfiguration,
  ctx: Context): SharedPreferences
}
```

PreferencesConfiguration 类代码如下所示。

```
package com.journaler.perferences
data class PreferencesConfiguration
(val key: String, val mode: Int)
```

PreferencesProvider 类代码如下所示。

```
package com.journaler.perferences

import android.content.Context
import android.content.SharedPreferences

class PreferencesProvider : PreferencesProviderAbstract() {
  override fun obtain(configuration: PreferencesConfiguration,
  ctx: Context): SharedPreferences {
    return ctx.getSharedPreferences(configuration.key,
    configuration.mode)
  }
}
```

可以看到，此处创建了一种简单机制，进而获取共享偏好设置。稍后将对其进行适当的整合。打开 MainActivity 类，并按照下列方式对其进行扩展。

```
class MainActivity : BaseActivity() {
  ...
  private val keyPagePosition = "keyPagePosition"
  ...
  override fun onCreate(savedInstanceState: Bundle?) {
    super.onCreate(savedInstanceState)
```

```
    val provider = PreferencesProvider()
    val config = PreferencesConfiguration("journaler_prefs",
    Context.MODE_PRIVATE)
    val preferences = provider.obtain(config, this)

    pager.adapter = ViewPagerAdapter(supportFragmentManager)
    pager.addOnPageChangeListener(object :
    ViewPager.OnPageChangeListener {
        override fun onPageScrollStateChanged(state: Int) {
          // Ignore
        }

        override fun onPageScrolled(position: Int, positionOffset:
        Float, positionOffsetPixels: Int) {
          // Ignore
        }
        override fun onPageSelected(position: Int) {
          Log.v(tag, "Page [ $position ]")
          preferences.edit().putInt(keyPagePosition,position).apply()
        }
     })

     val pagerPosition = preferences.getInt(keyPagePosition, 0)
     pager.setCurrentItem(pagerPosition, true)
     ...
   }
   ...
}
```

这里生成了 preferences 实例，用于持久化和读取视图分页器的位置。构建并运行应用程序，随后滑动至某一页面并关闭应用程序。接下来再次运行程序，当查看 Logcat 时，可看到下列输出内容（通过 Page 进行过滤）：

```
V/Main activity: Page [ 1 ]
V/Main activity: Page [ 2 ]
V/Main activity: Page [ 3 ]
After we restarted the application:
V/Main activity: Page [ 3 ]
V/Main activity: Page [ 2 ]
V/Main activity: Page [ 1 ]
V/Main activity: Page [ 0 ]
```

再次打开应用程序，并返回索引为 0 的页面。

8.4 本章小结

针对持久化应用程序的偏好设置，本章学习了 Android 共享偏好设置机制的应用方式。不难发现，其构建和使用过程均较为简单。第 9 章将重点考查 Android 中的并发机制。对此，我们将讨论 Android 中所提供的机制，并通过相关示例予以展示。

第 9 章　Android 中的并发机制

本章将讨论 Android 中的并发机制，并通过相关示例阐述如何在 Journaler 应用程序中使用并发操作。前述内容曾简单描述了 AsyncTask 类的一些基本知识，本章将对此加以深入讨论。

本章主要涉及以下主题：

- 处理程序和线程。
- AsyncTask。
- Android Looper。
- 延迟执行。

9.1　Android 并发机制简介

默认状态下，应用程序在主线程上被执行，同时，该应用程序必须是高性能的。如果某项任务的执行时间过长，那么将会得到一个 ANR，即 Android 应用程序没有响应消息。为了避免 ANR，可在后台运行代码。对此，Android 提供了相关机制，则可高效地完成此类任务。异步运行会对性能产生一定的影响，但会带来较好的用户体验。

所有的用户界面更新都将在某个线程上执行，该线程称为主线程。同时，所有的事件均在某个队列中被收集，并通过 Looper 类实例中被处理。

图 9.1 显示了所涉及类之间的关系。

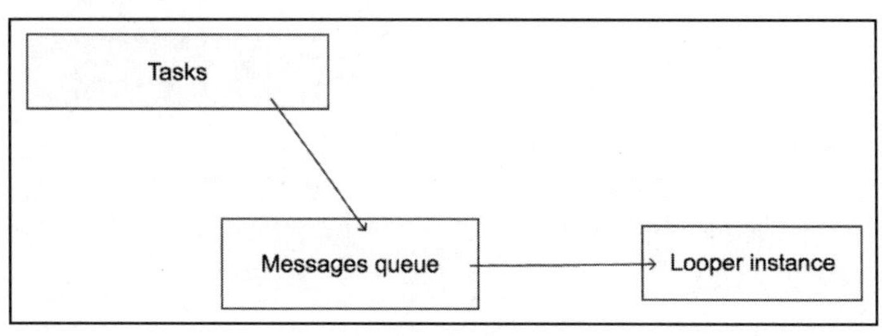

图 9.1

需要注意的是，主线程更新表示为全部可见 UI，但该过程可在另一个线程中完成。

直接从其他线程执行此操作将导致异常，且应用程序可能会崩溃。为了避免这种情况，可在主线程上执行所有与线程相关的代码，即在当前活动上下文中调用 runOnUiThread()方法。

9.2 处理程序和线程

在 Android 中，线程可通过标准方式被执行，但不建议在没有任何控制的情况下只触发裸线程。针对于此，可使用 ThreadPools 类和 Executor 类。

下面创建一个名为 execution 的新数据包，并定义一个 TaskExecutor 类，如下所示。

```
package com.journaler.execution

import java.util.concurrent.BlockingQueue
import java.util.concurrent.LinkedBlockingQueue
import java.util.concurrent.ThreadPoolExecutor
import java.util.concurrent.TimeUnit

class TaskExecutor private constructor(
    corePoolSize: Int,
    maximumPoolSize: Int,
    workQueue: BlockingQueue<Runnable>?

) : ThreadPoolExecutor(
    corePoolSize,
    maximumPoolSize,
    0L,
    TimeUnit.MILLISECONDS,
    workQueue
) {

companion object {
    fun getInstance(capacity: Int): TaskExecutor {
        return TaskExecutor(
          capacity,
          capacity * 2,
          LinkedBlockingQueue<Runnable>()
        )
    }
} }
```

此处，我们使用执行器实例化的成员方法扩展了 ThreadPoolExecutor 类和 companion 对象。下面将其应用于现有的代码上，并从使用的 AsyncTask 类切换到 TaskExecutor 类。对此，打开 NoteActivity 类，并按照下列方式对其进行更新：

```
class NoteActivity : ItemActivity() {
  ...
  private val executor = TaskExecutor.getInstance(1)
  ...
  private val locationListener = object : LocationListener {
    override fun onLocationChanged(p0: Location?) {
      p0?.let {
        LocationProvider.unsubscribe(this)
        location = p0
        val title = getNoteTitle()
        val content = getNoteContent()
        note = Note(title, content, p0)
        executor.execute {
          val param = note
          var result = false
          param?.let {
            result = Db.insert(param)
          }
          if (result) {
            Log.i(tag, "Note inserted.")
          } else {
              Log.e(tag, "Note not inserted.")
          }
        }
      }
    }

    override fun onStatusChanged(p0: String?, p1: Int, p2: Bundle?)
    {}
    override fun onProviderEnabled(p0: String?) {}
    override fun onProviderDisabled(p0: String?) {}
  }
  ...
  private fun updateNote() {
    if (note == null) {
      if (!TextUtils.isEmpty(getNoteTitle()) &&
      !TextUtils.isEmpty(getNoteContent())) {
        LocationProvider.subscribe(locationListener)
```

```
    }
  } else {
    note?.title = getNoteTitle()
    note?.message = getNoteContent()
    executor.execute {
      val param = note
      var result = false
      param?.let {
        result = Db.update(param)
      }
      if (result) {
        Log.i(tag, "Note updated.")
      } else {
        Log.e(tag, "Note not updated.")
      }
    }
  }
 }
... }
```

可以看到，此处利用执行器替换了 AsyncTask 类。该执行器一次仅处理单一线程。

除了标准的线程解决方案之外，Android 还提供了处理程序（handler）作为选项之一。处理程序并不是线程的替代品，而只是一项附加内容。处理程序实例利用其父线程注册自身，并体现了一种机制，进而可将数据发送至特定的线程中。相应地，我们可发送 Message 或 Runnable 类实例。下面通过示例阐述其应用方式。此处将利用一个指示器更新 Note 屏幕，其中，如果一切顺利执行，则指示器呈现为绿色；如果数据库持久化过程出现故障，则指示器将呈现为红色。另外，其默认的颜色为灰色。对此，打开 activity_note.xml 文件，并通过该指示器进行扩展。这里，指示器定义为一个视图，如下所示。

```
<?xml version="1.0" encoding="utf-8"?>
<ScrollView xmlns:android=
   "http://schemas.android.com/apk/res/android"
  android:layout_width="match_parent"
  android:layout_height="match_parent"
  android:fillViewport="true">

<LinearLayout
    android:layout_width="match_parent"
    android:layout_height="wrap_content"
    android:background="@color/black_transparent_40"
```

```
  android:orientation="vertical">

  ...

  <RelativeLayout
      android:layout_width="match_parent"
      android:layout_height="wrap_content">

      <View
          android:id="@+id/indicator"
          android:layout_width="40dp"
          android:layout_height="40dp"
          android:layout_alignParentEnd="true"
          android:layout_centerVertical="true"
          android:layout_margin="10dp"
          android:background="@android:color/darker_gray" />

      <EditText
          android:id="@+id/note_title"
          style="@style/edit_text_transparent"
          android:layout_width="match_parent"
          android:layout_height="wrap_content"
          android:hint="@string/title"
          android:padding="@dimen/form_padding" />

  </RelativeLayout>
   ...
 </LinearLayout>

</ScrollView>
```

当添加指示器时，取决于数据库插入结果，该指示器将会更改颜色。更新 NoteActivity 类，如下所示。

```
class NoteActivity : ItemActivity() {
  ...
 private var handler: Handler? = null
 ....
 override fun onCreate(savedInstanceState: Bundle?) {
   super.onCreate(savedInstanceState)
   handler = Handler(Looper.getMainLooper())
   ...
  }
```

```
...
private val locationListener = object : LocationListener {
  override fun onLocationChanged(p0: Location?) {
      p0?.let {
          ...
          executor.execute {
              ...
              handler?.post {
                   var color = R.color.vermilion
                   if (result) {
                       color = R.color.green
                   }
                   indicator.setBackgroundColor(
                           ContextCompat.getColor(
                                   this@NoteActivity,
                                   color
                           )
                   )
              }
          }
      }
  }

  override fun onStatusChanged(p0: String?, p1: Int, p2: Bundle?)
  {}
  override fun onProviderEnabled(p0: String?) {}
  override fun onProviderDisabled(p0: String?) {}
}
...
private fun updateNote() {
  if (note == null) {
      ...
  } else {
      ...
      executor.execute {
          ...
          handler?.post {
              var color = R.color.vermilion
              if (result) {
                  color = R.color.green
              }
              indicator.setBackgroundColor
```

```
            (ContextCompat.getColor(
                this@NoteActivity,
                color
            ))
        }
      }
    }
} }
```

构建、运行应用程序，并创建新的 Note。可以看到，在输入了标题和消息内容后，指示器的颜色将变为绿色。

稍作调整后将利用 Message 类实例执行相同的任务。根据下列内容更新代码：

```
class NoteActivity : ItemActivity() {
  ...
  override fun onCreate(savedInstanceState: Bundle?) {
    super.onCreate(savedInstanceState)
    handler = object : Handler(Looper.getMainLooper()) {
      override fun handleMessage(msg: Message?) {
        msg?.let {
          var color = R.color.vermilion
          if (msg.arg1 > 0) {
            color = R.color.green
          }
          indicator.setBackgroundColor
          (ContextCompat.getColor(
            this@NoteActivity,
            color
          ))
        }
        super.handleMessage(msg)
      }
    }
    ...
  }
  ...
  private val locationListener = object : LocationListener {
    override fun onLocationChanged(p0: Location?) {
      p0?.let {
        ...
        executor.execute {
          ...
          sendMessage(result)
```

```
          }
        }
      }

      override fun onStatusChanged(p0: String?, p1: Int, p2: Bundle?)
      {}
      override fun onProviderEnabled(p0: String?) {}
      override fun onProviderDisabled(p0: String?) {}
    }
    ...
   private fun updateNote() {
      if (note == null) {
          ...
      } else {
        ...
        executor.execute {
          ...
          sendMessage(result)
        }
      }
    }
    ...
   private fun sendMessage(result: Boolean) {
      val msg = handler?.obtainMessage()
      if (result) {
          msg?.arg1 = 1
      } else {
          msg?.arg1 = 0
      }
      handler?.sendMessage(msg)
    }
    ...
}
```

此处应注意 handler 实例和 sendMessage()方法。其中，我们利用 Handler 类中的 obtainMessage()方法获取 Message 实例。作为消息参数，这里传递了一个整数数据类型，并根据具体值更新指示器的颜色。

读者可能已经注意到，前述应用程序中曾使用了 AsyncTask 类，下面将在当前执行器上对其加以运行。

默认状态下，所有的 AsyncTask 均通过 Android 在队列中被执行。当以并行方式对其加以执行时，需要在执行器上完成这一工作。

当以并行方式执行多项任务时，假设当前任务数量为 2。相应地，任务执行完毕后将输出报告信息。随后，将任务数量增至 4。如果任务量并不繁重，工作依然可顺利完成。当前，可采用并行方式运行 50 个 AsyncTask。

由于对任务缺乏有效的控制，应用程序的执行速度将有所降低。对于这一类问题，需要适当管理任务以保持应有的性能。下面将继续对同一类进行调整，NoteActivity 类的修改方式如下所示。

```
class NoteActivity : ItemActivity() {
  ...
  private val threadPoolExecutor = ThreadPoolExecutor(
    3, 3, 1, TimeUnit.SECONDS, LinkedBlockingQueue<Runnable>()
  )

  private class TryAsync(val identifier: String) : AsyncTask<Unit,
    Int, Unit>() {
    private val tag = "TryAsync"

    override fun onPreExecute() {
      Log.i(tag, "onPreExecute [ $identifier ]")
      super.onPreExecute()
    }

    override fun doInBackground(vararg p0: Unit?): Unit {
      Log.i(tag, "doInBackground [ $identifier ][ START ]")
      Thread.sleep(5000)
      Log.i(tag, "doInBackground [ $identifier ][ END ]")
      return Unit
    }


    override fun onCancelled(result: Unit?) {
      Log.i(tag, "onCancelled [ $identifier ][ END ]")
      super.onCancelled(result)
    }

    override fun onProgressUpdate(vararg values: Int?) {
      val progress = values.first()
      progress?.let {
        Log.i(tag, "onProgressUpdate [ $identifier ][ $progress ]")
      }
      super.onProgressUpdate(*values)
```

```
    }

    override fun onPostExecute(result: Unit?) {
      Log.i(tag, "onPostExecute [ $identifier ]")
      super.onPostExecute(result)
    }
  }
  ...
  private val textWatcher = object : TextWatcher {
    override fun afterTextChanged(p0: Editable?) {
      ...
    }

    override fun beforeTextChanged(p0: CharSequence?, p1: Int, p2:
    Int, p3: Int) {}
    override fun onTextChanged(p0: CharSequence?, p1: Int, p2: Int,
    p3: Int) {
      p0?.let {
        tryAsync(p0.toString())
      }
    }
  }
  ...
  private fun tryAsync(identifier: String) {
    val tryAsync = TryAsync(identifier)
    tryAsync.executeOnExecutor(threadPoolExecutor)
  }
}
```

注意，上述代码仅做展示用，因而不可将其置入 Journaler 应用程序中。此处创建了新的 ThreadPoolExecutor 实例，对应的构造方法将接收以下参数：

- corePoolSize：表示为线程池中最小数量的线程。
- maximumPoolSize：表示为线程池中所允许的最大数量的线程。
- keepAliveTime：如果线程数量大于内核，则非内核线程将等待一项新的任务；如果未在该参数定义的时间范围内获得一项任务，则线程终止。
- Unit：表示为 keepAliveTime 的时间单位。
- WorkQueue：表示为用于加载任务的队列实例。
- 所有任务均在当前执行器上运行。AsyncTask 将在其生命周期内记录全部事件。在 main()方法中，等待的时间为 5 秒。运行应用程序并添加新的 Note（Android 作为标题），随后观察下列 Logcat 输出结果。

```
08-04 14:56:59.283 21953-21953...  I/TryAsync:onPreExecute [A]
08-04 14:56:59.284 21953-23233...  I/TryAsync: doInBackground [A][START]
08-04 14:57:00.202 21953-21953...  I/TryAsync:onPreExecute [An]
08-04 14:57:00.204 21953-23250 ... I/TryAsync: doInBackground [An][START]
08-04 14:57:00.783 21953-21953 ... I/TryAsync:onPreExecute [And]
08-04 14:57:00.784 21953-23281...  I/TryAsync:doInBackground [And][START]
08-04 14:57:01.001 21953-21953 ... I/TryAsync: onPreExecute [ Andr ]
08-04 14:57:01.669 21953-21953 ... I/TryAsync: onPreExecute [ Andro ]
08-04 14:57:01.934 21953-21953 ... I/TryAsync: onPreExecute [ Androi ]
08-04 14:57:02.314 21953-2195 ...  I/TryAsync: onPreExecute [ Android ]
08-04 14:57:04.285 21953-23233 ... I/TryAsync: doInBackground [ A ][ END ]
08-04 14:57:04.286 21953-23233 ... I/TryAsync:doInBackground [Andr][START]
08-04 14:57:04.286 21953-21953 ... I/TryAsync:onPostExecute [A]
08-04 14:57:05.204 21953-23250 ... I/TryAsync:doInBackground [An][END]
08-04 14:57:05.204 21953-21953 ... I/TryAsync:onPostExecute [An]
08-04 14:57:05.205 21953-23250 ... I/TryAsync:doInBackground [Andro][START]
08-04 14:57:05.784 21953-23281 ... I/TryAsync:doInBackground [And][END]
08-04 14:57:05.785 21953-23281 ... I/TryAsync:doInBackground [Androi][START]
08-04 14:57:05.786 21953-21953 ... I/TryAsync:onPostExecute [And]
08-04 14:57:09.286 21953-23233 ... I/TryAsync:doInBackground [Andr][END]
08-04 14:57:09.287 21953-21953 ... I/TryAsync:onPostExecute [Andr]
08-04 14:57:09.287 21953-23233 ... I/TryAsync:doInBackground [Android][START]
08-04 14:57:10.205 21953-23250 ... I/TryAsync:doInBackground [Andro][END]
08-04 14:57:10.206 21953-21953 ... I/TryAsync:onPostExecute [Andro]
08-04 14:57:10.786 21953-23281 ... I/TryAsync:doInBackground [Androi][END]
08-04 14:57:10.787 21953-2195 ...  I/TryAsync:onPostExecute [Androi]
08-04 14:57:14.288 21953-23233 ... I/TryAsync:doInBackground [Android][END]
08-04 14:57:14.290 21953-2195 ...  I/TryAsync:onPostExecute [Android]
```

下面通过任务中所执行的方法来过滤日志消息。首先，针对 onPreExecute()方法，查看下列过滤器：

```
08-04 14:56:59.283 21953-21953 ... I/TryAsync: onPreExecute [ A ]
08-04 14:57:00.202 21953-21953 ... I/TryAsync: onPreExecute [ An ]
08-04 14:57:00.783 21953-21953 ... I/TryAsync: onPreExecute [ And ]
08-04 14:57:01.001 21953-21953 ... I/TryAsync: onPreExecute [ Andr ]
08-04 14:57:01.669 21953-21953 ... I/TryAsync: onPreExecute [ Andro ]
08-04 14:57:01.934 21953-21953 ... I/TryAsync: onPreExecute [ Androi ]
08-04 14:57:02.314 21953-21953 ... I/TryAsync: onPreExecute [ Android ]
```

针对每个方法执行相同的操作，并重点关注方法被执行的时间。另外，还可修改 doInBackground()方法实现来执行一些更困难、更密集的工作。随后，通过输入较长的标题内容（例如完整的句子），以进一步引发多项任务。最后对日志内容进行过滤和分析。

9.3　理解 Android Looper

本节将讨论 Looper 类，该类在前述示例中曾有所介绍。

Looper 定义为一个类，并用于执行队列中的 Message 或 Runnable 实例。普通类一般不会包含 Looper 类中的任何队列。

这里的问题是，我们在何处可使用 Looper 类？当执行多个 Message 或 Runnable 实例时，Looper 类往往可派上用场。例如，在任务处理操作正运行时，则可向队列中添加一项新任务。

9.3.1　准备 Looper

当使用 Looper 类时，首先需要调用 prepare()方法。在准备 Looper 时，可使用 loop()方法，该方法用于在当前线程中创建 Message 循环，如下所示。

```
class LooperHandler : Handler() {
  override fun handleMessage(message: Message) {
      ...
  }
}

class LooperThread : Thread() {
  var handler: Handler? = null
  override fun run() {
    Looper.prepare()
    handler = LooperHandler()
    Looper.loop()
  }
}
```

上述示例展示了 Looper 类编程的基本步骤。这里，不要忘记对 Looper 类调用 prepare()方法；否则将会抛出异常，而应用程序将会随之崩溃。

9.3.2　延迟执行

本节将简要介绍 Android 中的延迟执行，同时还将展示一些应用于 UI 上的延迟操作示例。对此，打开 ItemsFragment 并进行如下调整：

```
class ItemsFragment : BaseFragment() {
```

```
 ...
   override fun onResume() {
     super.onResume()
     ...
     val items = view?.findViewById<ListView>(R.id.items)
     items?.let {
        items.postDelayed({
           if (!activity.isFinishing) {
              items.setBackgroundColor(R.color.grey_text_middle)
           }
        }, 3000)
     }
   }
  ...
}
```

3 秒钟后，如果未关闭此屏幕，背景颜色将被更改为略深的灰色调。运行应用程序并查看相应的结果。下面采用不同方式完成相同的任务，如下所示。

```
class ItemsFragment : BaseFragment() {
 ...
 override fun onResume() {
   super.onResume()
   ...
   val items = view?.findViewById<ListView>(R.id.items)
   items?.let {
       Handler().postDelayed({
           if (!activity.isFinishing) {
             items.setBackgroundColor(R.color.grey_text_middle)
           }
       }, 3000)
    }
  }
  ...
}
```

此处使用了 Handler 类型执行延迟修改。

9.4 本章小结

本章介绍了 Android 的并发机制，并通过相关示例予以展示，以便为深入理解 Android 服务打下良好的基础。Android 服务是一种功能强大的并发特性，并可视为应用程序的核心内容。

第 10 章　Android 服务

第 9 章讨论了 Android 中的并发机制，但事情远未结束。本章将讨论 Android Framework 中最重要的部分之一，即 Android 服务，包括其概念、应用时机和方式。

本章主要涉及以下主题：

- 服务分类。
- Android 服务的基础知识。
- 定义主应用程序服务。
- 定义意图（Intent）服务。

10.1　服务分类

在定义 Android 服务分类并深入了解每种类型之前，首先需要明晰 Android 服务的具体含义。Android 服务是 Android Framework 提供的一种机制，据此，可将运行时间较长的任务移至后台。Android 服务内含了一些较好的附加特性，进而在提升灵活性的同时简化开发工作。对此，接下来将通过扩展 Journaler 应用程序创建一项服务。

Android 服务可视为一个应用程序组件，且不包含任何 UI；另外，Android 服务可通过 Android 应用程序启动，必要时可持续处于运行状态，甚至是离开或终止当前应用程序。

Android 服务主要涵盖以下 3 种类型：

- 前台服务。
- 后台服务。
- 绑定服务。

10.1.1　Android 前台服务

前台服务执行终端用户可以察觉到的任务，此类服务需要显示一个状态栏图标。即使未与应用程序交互，它们也会处于持续运行状态。

10.1.2　Android 后台服务

与前台服务不同，后台服务执行任务时不会被终端用户察觉。例如，可与后台实例执行同步操作，且用户无须了解实际的处理过程。因此，我们决定不去打扰终端用户，所有操作都将在应用程序的后台无声地执行。

10.1.3　Android 绑定服务

应用程序组件可绑定至某项服务上，并触发执行不同的任务。与 Android 中的服务交互十分简单，当组件绑定至某项服务上后，只要存在一个此类组件，该服务即会处于运行状态；如果服务上不存在任何绑定组件，则该服务被销毁。

因此，可创建一项后台服务（于后台运行），并可对其进行绑定。

10.2　Android 服务基础知识

当定义 Android 服务时，需要扩展 Service 类，并重载以下方法：

- ❑ onStartCommand()：当 startService()方法被某个 Android 组件触发后，该方法将被执行，随后将启动 Android 服务并于后台运行；当终止该服务时，须执行与 startService()方法功能相反的 stopService()方法。
- ❑ onBind()：当从另一个 Android 组件绑定服务时，可使用 bindService()方法，随后执行 onBind()方法。在该方法的服务实现中，须通过返回 IBinder 类实例以提供客户端与服务通信的接口。实现该方法并非可选项，但如果不打算绑定至服务，仅返回 null 即可。
- ❑ onCreate()：当创建服务时将执行该方法。如果服务已处于运行状态，则不执行该方法。
- ❑ onDestroy()：当服务被销毁时，执行该方法。此处可针对具体服务重载该方法，并执行全部的清空任务。
- ❑ onUnbind()：当从服务中接触绑定时将执行该方法。

10.2.1　声明服务

当声明服务时，需要向 Android Manifest 中添加对应类。下列代码片段解释了 Android Manifest 中服务的定义方式：

```
<manifest xmlns:android=
 "http://schemas.android.com/apk/res/android"
```

```
  package="com.journaler">
  ...
  <application ... >
    <service
      android:name=".service.MainService"
      android:exported="false" />
      ...
  </application>
</manifest>
```

可以看到，此处定义了 MainService 类，该类扩展了 Service 类且位于 service 数据包中。其中，exported 标记设置为 false，这意味着，service 将运行于与当前应用程序相同的进程中。相应地，若在独立进程中运行 service 时，可将该标记设置为 true。

需要注意的是，Service 类不是唯一可以扩展的类。除此之外，IntentService 类也是可用的。这里，IntentService 类表示为继承自 Service 的类，并通过 worker 线程逐一处理请求。对此，需要实现 onHandleIntent()方法。IntentService 类扩展如下：

```
public class MainIntentService extends IntentService {
  /**
  * A constructor is mandatory!
  */
  public MainIntentService() {
    super("MainIntentService");
  }

  /**
  * All important work is performed here.
  */
  @Override
  protected void onHandleIntent(Intent intent) {
    // Your implementation for handling received intents.
  }
}
```

下面继续讨论 Service 类的扩展，并通过下列方式重载 onStartCommand()方法：

```
override fun onStartCommand(intent: Intent?, flags: Int, startId:
Int): Int {
  return Service.START_STICKY
}
```

这里的问题是，START_STICKY 返回结果的具体含义是什么？如果服务被系统终止，或者终止了服务所属的应用程序，那么，服务将重新启动。与此相对的是 START_NOT_STICKY，其中，服务将不会被重新创建和重新启动。

10.2.2　启动服务

当启动服务时，需要定义代表该服务的意图（Intent）。下列代码展示了如何启动服务：

```
val startServiceIntent = Intent(ctx, MainService::class.java)
ctx.startService(startServiceIntent)
```

其中，ctx 表示为 Android Context 类的有效实例。

10.2.3　终止服务

当终止服务时，可执行 Android Context 类中的 stopService()方法，如下所示。

```
val stopServiceIntent = Intent(ctx, MainService::class.java)
ctx.stopService(startServiceIntent)
```

10.2.4　绑定 Android 服务

当执行服务绑定时，需要调用 bindService()方法。当与源自活动或其他 Android 组件的服务进行交互时，服务绑定不可或缺。在绑定过程中，需要实现 onBind()方法，并返回一个 IBinder 实例。如果不存在任何参与者，并且所有这些参与者都是未绑定的，Android 就会销毁该服务。

10.2.5　终止服务

如前所述，stopService 将终止服务。在我们的服务实现中，也可调用 stopSelf()方法完成这一任务。

10.2.6　服务的生命周期

前述内容介绍了 Android 服务生命周期中所执行的所有重要方法。与其他 Android 组件类似，服务包含其自身的生命周期。图 10.1 对此进行了总结。

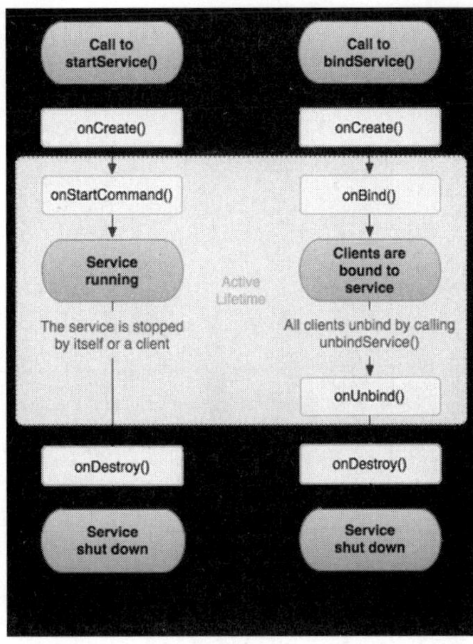

图 10.1

在理解了 Android 服务后，下面将构建自己的服务，并扩展 Journaler 应用程序。在后续章节中，还将会进一步丰富该服务的内容。

10.3　定义主应用程序服务

如前所述，当前应用程序将处理 Note 和 Todo，相关数据将通过本地方式存储于 SQLite 数据库中，且与运行于远程服务器上的后台实例保持同步。全部与同步相关的操作均在应用程序的后台中执行。相应地，全部职责均由当前服务所承担，接下来将对该服务加以定义。对此，创建一个名为 service 的数据包，以及一个名为 MainService 的新类，该类将扩展 Android 的 Service 类，如下所示。

```
class MainService : Service(), DataSynchronization {

 private val tag = "Main service"
 private var binder = getServiceBinder()
 private var executor = TaskExecutor.getInstance(1)

 override fun onCreate() {
   super.onCreate()
   Log.v(tag, "[ ON CREATE ]")
```

```
}

override fun onStartCommand(intent: Intent?, flags: Int, startId:
Int): Int {
  Log.v(tag, "[ ON START COMMAND ]")
  synchronize()
  return Service.START_STICKY
}

override fun onBind(p0: Intent?): IBinder {
  Log.v(tag, "[ ON BIND ]")
  return binder
}

override fun onUnbind(intent: Intent?): Boolean {
  val result = super.onUnbind(intent)
  Log.v(tag, "[ ON UNBIND ]")
  return result
}

override fun onDestroy() {
  synchronize()
  super.onDestroy()
  Log.v(tag, "[ ON DESTROY ]")
}

override fun onLowMemory() {
  super.onLowMemory()
  Log.w(tag, "[ ON LOW MEMORY ]")
}

override fun synchronize() {
  executor.execute {
      Log.i(tag, "Synchronizing data [ START ]")
      // For now we will only simulate this operation!
      Thread.sleep(3000)
      Log.i(tag, "Synchronizing data [ END ]")
  }
}

private fun getServiceBinder(): MainServiceBinder =
MainServiceBinder()

inner class MainServiceBinder : Binder() {
```

```
    fun getService(): MainService = this@MainService
  }
}
```

此处扩展了 Android 的 Service 类，进而获取全部服务功能。除此之外，我们还实现了描述服务主要功能的 DataSynchronization 接口（具有同步特征），如下所示。

```
package com.journaler.service
interface DataSynchronization {
  fun synchronize()
}
```

因此，我们定义了 synchronize()方法的实现，进而模拟实际的同步操作。稍后，还将更新上述代码以实现真正的后台通信。

相应地，所有重要的生命周期方法均被重载。此处应注意 bind()方法，该方法返回一个通过调用 getServiceBinder()方法所生成的绑定器实例。基于 MainServiceBinder 类，我们将向终端用户公开 service 实例，进而在必要时触发同步机制。

除了终端用户之外，同步行为还可通过服务自身被触发。相应地，当服务启动和销毁时，我们将触发同步操作。

下面考查另一个较为重要的问题，即启动和终止 MainService 类。对此，打开体现当前应用程序的 Journaler 类，并进行如下更新：

```
class Journaler : Application() {

  companion object {
    val tag = "Journaler"
    var ctx: Context? = null
  }

  override fun onCreate() {
    super.onCreate()
    ctx = applicationContext
    Log.v(tag, "[ ON CREATE ]")
    startService()
  }

  override fun onLowMemory() {
    super.onLowMemory()
    Log.w(tag, "[ ON LOW MEMORY ]")
    // If we get low on memory we will stop service if running.
    stopService()
  }
```

```
  override fun onTrimMemory(level: Int) {
    super.onTrimMemory(level)
    Log.d(tag, "[ ON TRIM MEMORY ]: $level")
  }

  private fun startService() {
    val serviceIntent = Intent(this, MainService::class.java)
    startService(serviceIntent)
  }

  private fun stopService() {
    val serviceIntent = Intent(this, MainService::class.java)
    stopService(serviceIntent)
  }

}
```

当创建 Journaler 应用程序时，将启动 MainService 类。同时，此处还实现了一个优化操作——若内存空间较少，将终止 MainService 类。由于服务启动时是黏性的，因此，如果显式地终止应用程序，服务将重新启动。

截至目前，我们已经讨论了服务的启动、终止及其实现过程。读者可能还记得之前的模型，在应用程序抽屉菜单的底部计划再添加一个条目，并设置 synchronize 按钮。触发该按钮将执行与后台的同步操作。

接下来将添加抽屉菜单条目并将其与当前服务连接。对此，打开 NavigationDrawerItem 类，并按照下列方式进行更新：

```
data class NavigationDrawerItem(
  val title: String,
  val onClick: Runnable,
  var enabled: Boolean = true
)
```

此处引入了 enabled 参数。与此类似，必要时，应用程序抽屉菜单条目也可被禁用。默认状态下，synchronize 按钮将处于禁用状态；当与 main 服务绑定时，该按钮将处于启用状态。此类变化也会影响 NavigationDrawerAdapter。对此，可参考下列代码：

```
class NavigationDrawerAdapter(
  val ctx: Context,
  val items: List<NavigationDrawerItem>
  ) : BaseAdapter() {
```

```
  private val tag = "Nav. drw. adptr."

  override fun getView(position: Int, v: View?, group:
  ViewGroup?): View {
    ...
    val item = items[position]
    val title = view.findViewById<Button>(R.id.drawer_item)
    ...
    title.setOnClickListener {
      if (item.enabled) {
          item.onClick.run()
      } else {
          Log.w(tag, "Item is disabled: $item")
      }
    }

    return view
  }
  ...
}
```

最后，还需要更新 MainActivity 类，以使 synchronize 按钮可触发同步操作，如下所示。

```
class MainActivity : BaseActivity() {
  ...
  private var service: MainService? = null
  private val synchronize: NavigationDrawerItem by lazy {
    NavigationDrawerItem(
      getString(R.string.synchronize),
      Runnable { service?.synchronize() },
      false
    )
  }

  private val serviceConnection = object : ServiceConnection {
    override fun onServiceDisconnected(p0: ComponentName?) {
        service = null
        synchronize.enabled = false
    }

    override fun onServiceConnected(p0: ComponentName?, binder:
    IBinder?) {
      if (binder is MainService.MainServiceBinder) {
        service = binder.getService()
```

```
          service?.let {
            synchronize.enabled = true
          }
        }
      }
    }

    override fun onCreate(savedInstanceState: Bundle?) {
      super.onCreate(savedInstanceState)
      ...
      val menuItems = mutableListOf<NavigationDrawerItem>()
      ...
      menuItems.add(synchronize)
      ...
    }

    override fun onResume() {
      super.onResume()
      val intent = Intent(this, MainService::class.java)
      bindService(intent, serviceConnection,
      android.content.Context.BIND_AUTO_CREATE)
    }

    override fun onPause() {
      super.onPause()
      unbindService(serviceConnection)
    }
    ...
}
```

无论主活动是否处于活动状态，我们都将绑定或取消绑定 main 服务。当执行绑定时，需要使用 ServiceConnection 实现，因为它将根据绑定状态启用或禁用同步按钮。另外，还将根据绑定状态维护 main 服务实例。同步按钮将访问 service 实例，并在单击时触发 synchronize()方法。

10.4　定义 Intent 服务

当前，main 服务处于运行状态，并定义了相关职责，下面将引入多项服务进行改进当前应用程序。此时，我们将定义 Intent 服务。Intent 服务将负责执行数据库 CRUD 操作。基本上讲，这里将定义 Intent 服务并对已有的代码进行重构。

首先，可在 service 数据包中创建名为 DatabaseService 的新类。在实现该类之前，需要将其注册至 Android Manifest 中，如下所示。

```
<manifest xmlns:android=
  "http://schemas.android.com/apk/res/android"
   package="com.journaler">
   ...
  <application ... >
  <service
    android:name=".service.MainService"
    android:exported="false" />

  <service
    android:name=".service.DatabaseService"
    android:exported="false" />
    ...
  </application>
</manifest>

Define DatabaseService like this:
class DatabaseService :
 IntentService("DatabaseService") {
   companion object {
     val EXTRA_ENTRY = "entry"
     val EXTRA_OPERATION = "operation"
   }

   private val tag = "Database service"

   override fun onCreate() {
     super.onCreate()
     Log.v(tag, "[ ON CREATE ]")
   }

   override fun onLowMemory() {
     super.onLowMemory()
     Log.w(tag, "[ ON LOW MEMORY ]")
   }

   override fun onDestroy() {
     super.onDestroy()
     Log.v(tag, "[ ON DESTROY ]")
   }
```

```
    override fun onHandleIntent(p0: Intent?) {
      p0?.let {
        val note = p0.getParcelableExtra<Note>(EXTRA_ENTRY)
        note?.let {
          val operation = p0.getIntExtra(EXTRA_OPERATION, -1)
          when (operation) {
            MODE.CREATE.mode -> {
              val result = Db.insert(note)
              if (result) {
                Log.i(tag, "Note inserted.")
              } else {
                Log.e(tag, "Note not inserted.")
                }
              }
              MODE.EDIT.mode -> {
                val result = Db.update(note)
                if (result) {
                  Log.i(tag, "Note updated.")
                } else {
                  Log.e(tag, "Note not updated.")
                }
              }
              else -> {
                  Log.w(tag, "Unknown mode [ $operation ]")
              }
            }
          }
        }
    }

}
```

main 服务将接收 Intent、获取对应操作并从中标注实例。取决于具体的操作，这将会触发相应的 CRUD 操作。当向 Intent 传递一个 Note 实例时，需要实现 Parcelable，以便数据可被高效地传递。与 Serializable 相比，Parcelable 将更加快速。因此，需要对代码进行适当的优化。这里将执行显式的序列化操作，且不使用反射。打开 Note 类，并按照下列方式进行更新：

```
package com.journaler.model
import android.location.Location
import android.os.Parcel
```

```
import android.os.Parcelable

class Note(
  title: String,
  message: String,
  location: Location
) : Entry(
  title,
  message,
  location
), Parcelable {

  override var id = 0L

  constructor(parcel: Parcel) : this(
    parcel.readString(),
    parcel.readString(),
    parcel.readParcelable(Location::class.java.classLoader)
  ) {
    id = parcel.readLong()
  }

  override fun writeToParcel(parcel: Parcel, flags: Int) {
    parcel.writeString(title)
    parcel.writeString(message)
    parcel.writeParcelable(location, 0)
    parcel.writeLong(id)
  }

  override fun describeContents(): Int {
    return 0
  }

  companion object CREATOR : Parcelable.Creator<Note> {
    override fun createFromParcel(parcel: Parcel): Note {
      return Note(parcel)
    }

    override fun newArray(size: Int): Array<Note?> {
      return arrayOfNulls(size)
    }
  }

}
```

当通过 Intent 传递至 DatabaseService 时，Note 类将被高效地序列化和反序列化。

最后一部分是对当前的 CRUD 操作进行调整。此处并不直接从 NoteActivity 类中访问 Db 类，而是创建一个 Intent 并启动它，以便当前服务为我们处理其余的工作。

打开 NoteActivity 类，并按照下列方式更新代码：

```
class NoteActivity : ItemActivity() {
  ...
  private val locationListener = object : LocationListener {
    override fun onLocationChanged(p0: Location?) {
      p0?.let {
        LocationProvider.unsubscribe(this)
        location = p0
        val title = getNoteTitle()
        val content = getNoteContent()
        note = Note(title, content, p0)

        // Switching to intent service.
        val dbIntent = Intent(this@NoteActivity,
        DatabaseService::class.java)
        dbIntent.putExtra(DatabaseService.EXTRA_ENTRY, note)
        dbIntent.putExtra(DatabaseService.EXTRA_OPERATION,
        MODE.CREATE.mode)
        startService(dbIntent)
        sendMessage(true)
      }
    }

  override fun onStatusChanged(p0: String?, p1: Int, p2: Bundle?) {}
  override fun onProviderEnabled(p0: String?) {}
  override fun onProviderDisabled(p0: String?) {}
}
...
private fun updateNote() {
  if (note == null) {
    if (!TextUtils.isEmpty(getNoteTitle()) &&
    !TextUtils.isEmpty(getNoteContent())) {
      LocationProvider.subscribe(locationListener)
    }
    } else {
      note?.title = getNoteTitle()
      note?.message = getNoteContent()
```

```
    // Switching to intent service.
    val dbIntent = Intent(this@NoteActivity,
    DatabaseService::class.java)
    dbIntent.putExtra(DatabaseService.EXTRA_ENTRY, note)
    dbIntent.putExtra(DatabaseService.EXTRA_OPERATION,
    MODE.EDIT.mode)
    startService(dbIntent)
    sendMessage(true)
   }
  }
  ...
}
```

不难发现，修改过程十分简单。随后，构建并运行应用程序。当创建或更新 Note 类时，应留意与执行数据库操作相关的日志信息。另外，还应注意日志中与 DatabaseService 生命周期方法相关的内容。

10.5　本 章 小 结

本章解释了 Android 服务的具体含义，以及 Android 服务的各种类型，并展示了相应的应用示例。在实现过程中，也建议读者至少再考查一项服务。该服务可涵盖应用程序的某些已有部分，或者引入全新的内容，进而查看此类方案的优点。

第 11 章　消 息 机 制

本章将与 Android 广播协同工作，将其用作一种机制接收和发送消息，其间将会涉及多个步骤。本章首先将深入讨论这一机制，以及如何应用 Android 广播消息；随后将监听一些较为常见的消息，并创建新消息进而对其加以广播；最后，本章还将介绍消息的启用、关闭和网络传输，因此，相关应用程序将主要关注这几方面的系统事件。

本章主要涉及以下主题：

- Android 广播。
- 监听广播。
- 创建广播。
- 监听网络事件。

11.1　理解 Android 广播

Android 应用程序可发送或接收消息。相应地，消息可以是与系统相关的事件，或者是自定义事件。通过定义适当的意图过滤器和广播接收器，相关方将针对特定的消息加以注册。当消息处于广播状态，所有关注的各方将会被通知。需要注意的是，一旦订阅了广播消息（尤其是 Activity 类），还需要在某个时间点取消订阅。这里的问题是，何时可使用广播消息？对此，当应用程序间需要使用一个消息传输系统时，即可使用广播消息。例如在后台启动了一个运行时间较长的进程。在某个时间点，需要通知多个与处理结果相关的上下文。因此，广播消息是一种较好的解决方案。

11.1.1　系统广播

系统广播是指，当产生各种系统事件时，由 Android 系统发送的广播。所发送和最终接收的消息封装于 Intent 类中，该类包含了与特定事件相关的信息。另外，每个 Intent 类须包含相应的动作集，例如 android.intent.action.ACTION_POWER_CONNECTED。相应地，与事件相关的信息通过绑定的附加数据表示——可绑定一个附加的字符串字段以表示与所关注事件相关的特定数据。下面考查一个充电和电量信息的示例。每次电池状态发生变化时，使用者将被通知，并接收包含电池电量信息的广播消息，如下所示。

```
val intentFilter = IntentFilter(Intent.ACTION_BATTERY_CHANGED)
val batteryStatus = registerReceiver(null, intentFilter)
val status = batteryStatus.getIntExtra(BatteryManager.EXTRA_STATUS, -1)

val isCharging =
          status == BatteryManager.BATTERY_STATUS_CHARGING ||
          status == BatteryManager.BATTERY_STATUS_FULL

val chargePlug = batteryStatus.getIntExtra(BatteryManager.
 EXTRA_PLUGGED, -1)
val usbCharge = chargePlug == BatteryManager.BATTERY_PLUGGED_USB
val acCharge = chargePlug == BatteryManager.BATTERY_PLUGGED_AC
```

在上述示例中，针对电池信息注册了 Intent 过滤器，但此处并未传递一个广播接收器实例，其原因在于，电池数据具有“黏性”特征。这里，黏性 Intent 是指，在广播执行后一段时间内仍存在的 Intent。注册此类数据将即刻返回包含最新数据的 Intent。下面将传递一个广播接收器实例：

```
val receiver = object : BroadcastReceiver() {
  override fun onReceive(p0: Context?, batteryStatus: Intent?) {
  val status = batteryStatus?.getIntExtra
  (BatteryManager.EXTRA_STATUS, -1)
            val isCharging =
                  status ==
                  BatteryManager.BATTERY_STATUS_CHARGING ||
                  status == BatteryManager.BATTERY_STATUS_FULL
                  val chargePlug = batteryStatus?.getIntExtra
                  (BatteryManager.EXTRA_PLUGGED, -1)
                  val usbCharge = chargePlug ==
                  BatteryManager.BATTERY_PLUGGED_USB
                  val acCharge = chargePlug ==
                  BatteryManager.BATTERY_PLUGGED_AC
  }
}

val intentFilter = IntentFilter(Intent.ACTION_BATTERY_CHANGED)
registerReceiver(receiver, intentFilter)
```

每次电池信息发生变化时，接收器将执行其实现中定义的代码；此外，还可在 Android Manifest 中定义自己的接收器，如下所示。

```
<receiver android:name=".OurPowerReceiver">
```

```
  <intent-filter>
    <action android:name="android.intent.action.
    ACTION_POWER_CONNECTED"/>
    <action android:name="android.intent.action.
    ACTION_POWER_DISCONNECTED"/>
  </intent-filter>
</receiver>
```

11.1.2　监听广播

在前述示例中曾提及，可通过以下两种方式之一接收广播：

- ❑ 通过 Android Manifest 注册广播接收器。
- ❑ 利用上下文中的 registerBroadcast()方法注册广播。

Manifest 声明须满足以下内容：

- ❑ 包含 android:name 和 android:exported 参数的<receiver>元素。
- ❑ 针对所订阅的动作，接收器必须包含 Intent 过滤器，考查下列示例：

```
<receiver android:name=".OurBootReceiver"
  android:exported="true">
  <intent-filter>
    <action android:name=
    "android.intent.action.BOOT_COMPLETED"/>
    ...
    <action android:name="..."/>
    <action android:name="..."/>
    <action android:name="..."/>
  </intent-filter>

</receiver>
```

可以看到，name 表示为广播接收器类的名称。其中，exported 表示为应用程序是否可从接收器应用程序的外部资源中接收消息。

BroadcastReceiver 的子类如下所示。

```
val receiver = object : BroadcastReceiver() {
   override fun onReceive(ctx: Context?, intent: Intent?) {
       // Handle your received code.
   }
}
```

注意，onReceive()方法中所执行的操作不应占用太多时间，否则将会产生 ANR。

11.1.3　从上下文中注册

本节将通过示例展示从 Android 上下文中注册广播接收器。对此，需要使用一个接收器实例，假设该实例为 myReceiver，如下所示。

```
val myReceiver = object : BroadcastReceiver(){
  ...

  }We need intent filter prepared:
  val filter =
  IntentFilter(ConnectivityManager.CONNECTIVITY_ACTION)
  registerReceiver(myReceiver, filter)
```

上述示例将注册一个监听连接信息的接收器。由于该接收器从当前上下文中进行注册，因而仅当所注册的上下文是有效的，该接收器方为有效。除此之外，还可使用 LocalBroadcastManager 类。LocalBroadcastManager 的目的是注册 Intent 广播并将其发送到进程中的本地对象，如下所示。

```
LocalBroadcastManager
  .getInstance(applicationContext)
  .registerReceiver(myReceiver, intentFilter)
```

当解除注册时，可执行下列代码片段：

```
LocalBroadcastManager
  .getInstance(applicationContext)
  .unregisterReceiver(myReceiver)
```

对于上下文订阅的接收器，应注意其注册的解除操作。例如，如果在活动的 onCreate()方法中注册一个接收器，则必须在 onDestroy()方法中对其进行解除；否则将会出现接收器泄露问题。类似地，如果在活动的 onResume()方法中进行注册，则必须在 onPause()方法中解除注册；否则将会注册多次。

11.1.4　接收器的执行

在 onReceive()方法实现中执行的代码应视为前台进程。广播接收器在从该方法中返回之前一直处于活动状态。相应地，系统将运行定义于该实现中的代码，除非内存空间严重不足。如前所述，此处仅应执行相对简短的操作，否则将会出现 ANR 问题。当接收消息且执行较为耗时的操作时，一种较好的方法是启动 AsyncTask，并于其中执行全部工作，如下所示。

```
class AsyncReceiver : BroadcastReceiver() {
  override fun onReceive(p0: Context?, p1: Intent?) {
    val pending = goAsync()
    val async = object : AsyncTask<Unit, Unit, Unit>() {
      override fun doInBackground(vararg p0: Unit?) {
        // Do some intensive work here...
        pending.finish()
      }
    }
   async.execute()
  }
}
```

上述示例中引入了 goAsync()方法应用。该方法返回 PendingResult 类型对象，表示一个从 API 方法调用的待定对象。在调用当前实例的 finish()方法之前，Android 系统视为接收器处于活动状态。通过这种机制，可在广播接收器中执行异步处理。在结束了密集型工作之后，可调用 finish()方法向 Android 系统表明该组件可以回收。

11.1.5 发送广播

Android 包含以下 3 种方式发送广播消息：

- 使用 sendOrderedBroadcast(Intent, String)方法一次向一个接收器发送消息。由于接收器按顺序执行，因而可将某个结果传播至下一个接收器。另外，还可中止广播，这样广播就不会传递给其他接收器。相应地，可控制接收器的执行顺序。对此，可以使用匹配 Intent 过滤器的 android:priority 属性来确定优先级。
- 使用 sendBroadcast(Intent)方法向所有的接收器传送广播消息。该传送过程处于无序状态。
- 使用 LocalBroadcastManager.sendBroadcast(Intent)方法将广播发送给与发送方位于同一应用程序中的接收方。

下列代码将广播消息发送至所有的关注方：

```
val intent = Intent()
intent.action = "com.journaler.broadcast.TODO_CREATED"
intent.putExtra("title", "Go, buy some lunch.")
intent.putExtra("message", "For lunch we have chicken.")
sendBroadcast(intent)
```

此处创建了包含了附加数据（与 note 相关，即标题和消息）的广播消息。针对当前动作，所有的关注方需要使用一个相应的 IntentFilter 实例，如下所示。

```
com.journaler.broadcast.TODO_CREATED
```

这里不应将启动活动与发送广播消息予以混淆。Intent 类仅用于当前信息的封装器，这两个操作彼此间完全不同。另外，通过本地广播机制，可实现相同的任务，如下所示。

```
val ctx = ...
val broadcastManager = LocalBroadcastManager.getInstance(ctx)
val intent = Intent()
intent.action = "com.journaler.broadcast.TODO_CREATED"
intent.putExtra("title", "Go, buy some lunch.")
intent.putExtra("message", "For lunch we have chicken.")
broadcastManager.sendBroadcast(intent)
```

接下来将考查广播消息机制中最为重要的部分，并对应用程序实施进一步的扩展。具体来说，Journaler 程序将发送和接收包含数据的自定义广播消息，并与系统广播进行交互，例如系统启动、关闭和网络。

11.2　创建自己的广播消息

前述内容曾对 NoteActivity 类进行了重构，接下来将展示重要内容中的最后一个状态。

```
class NoteActivity : ItemActivity() {
  ...
  private val locationListener = object : LocationListener {
    override fun onLocationChanged(p0: Location?) {
      p0?.let {
        LocationProvider.unsubscribe(this)
        location = p0
        val title = getNoteTitle()
        val content = getNoteContent()
        note = Note(title, content, p0)

        // Switching to Intent service.
        val dbIntent = Intent(this@NoteActivity,
        DatabaseService::class.java)
        dbIntent.putExtra(DatabaseService.EXTRA_ENTRY, note)
        dbIntent.putExtra(DatabaseService.EXTRA_OPERATION,
        MODE.CREATE.mode)
```

```
      startService(dbIntent)
      sendMessage(true)
    }
  }

  override fun onStatusChanged(p0: String?, p1: Int, p2: Bundle?) {}
  override fun onProviderEnabled(p0: String?) {}
  override fun onProviderDisabled(p0: String?) {}
 }
 ...
 private fun updateNote() {
  if (note == null) {
    if (!TextUtils.isEmpty(getNoteTitle()) &&
      !TextUtils.isEmpty(getNoteContent())) {
       LocationProvider.subscribe(locationListener)
    }
  } else {
    note?.title = getNoteTitle()
    note?.message = getNoteContent()

    // Switching to Intent service.
    val dbIntent = Intent(this@NoteActivity,
    DatabaseService::class.java)
    dbIntent.putExtra(DatabaseService.EXTRA_ENTRY, note)
    dbIntent.putExtra(DatabaseService.EXTRA_OPERATION, MODE.EDIT.mode)
    startService(dbIntent)
    sendMessage(true)
  }
 }
 ...
}
```

可以看到，这里向执行服务发送了 Intent，但由于尚未获取返回值，因而仅执行了包含布尔值 true 作为参数的 sendMessage()方法。此处使用了一个期望值以表示 CRUD 的操作结果，即成功或失败。我们将使用广播消息将服务与 NoteActivity 连接起来。每次插入或更新 Note 广播时，将会引发一条消息。于 NoteActivity 中定义的监听器将响应于该消息，并将触发 sendMessage()方法。下面将对代码进行更新。对此，打开 Crud 接口，并使用包含动作常量和 CRUD 操作结果的伴生对象对其进行扩展，如下所示。

```
interface Crud<T> {
 companion object {
```

```
    val BROADCAST_ACTION = "com.journaler.broadcast.crud"
    val BROADCAST_EXTRAS_KEY_CRUD_OPERATION_RESULT = "crud_result"
  }
  ...
}
```

接下来打开 DatabaseService，并通过下列方法进行扩展，该方法负责在 CRUD 操作执行上发送广播消息。

```
class DatabaseService : IntentService("DatabaseService") {
  ...
  override fun onHandleIntent(p0: Intent?) {
    p0?.let {
      val note = p0.getParcelableExtra<Note>(EXTRA_ENTRY)
      note?.let {
          val operation = p0.getIntExtra(EXTRA_OPERATION, -1)
          when (operation) {
              MODE.CREATE.mode -> {
                  val result = Db.insert(note)
                  if (result) {
                      Log.i(tag, "Note inserted.")
                  } else {
                      Log.e(tag, "Note not inserted.")
                  }
                  broadcastResult(result)
              }
              MODE.EDIT.mode -> {
                  val result = Db.update(note)
                  if (result) {
                      Log.i(tag, "Note updated.")
                  } else {
                      Log.e(tag, "Note not updated.")
                  }
                  broadcastResult(result)
              }
              else -> {
                  Log.w(tag, "Unknown mode [ $operation ]")
              }
          }
      }
    }
  }
  ...
```

```
  private fun broadcastResult(result: Boolean) {
    val intent = Intent()
    intent.putExtra(
          Crud.BROADCAST_EXTRAS_KEY_CRUD_OPERATION_RESULT,
          if (result) {
             1
          } else {
             0
          }
    )
} }
```

其中，我们得到了 CRUD 操作结果，并将其作为消息进行广播。同时，NoteActivity 将对此进行监听，如下所示。

```
class NoteActivity : ItemActivity() {
  ...
  private val crudOperationListener = object : BroadcastReceiver() {
      override fun onReceive(ctx: Context?, intent: Intent?) {
          intent?.let {
              val crudResultValue =
              intent.getIntExtra(MODE.EXTRAS_KEY, 0)
              sendMessage(crudResultValue == 1)
          }
      }
  }
  ...
  override fun onCreate(savedInstanceState: Bundle?) {
      ....
      registerReceiver(crudOperationListener, intentFiler)
  }

  override fun onDestroy() {
      unregisterReceiver(crudOperationListener)
      super.onDestroy()
  }
  ...
  private fun sendMessage(result: Boolean) {
      Log.v(tag, "Crud operation result [ $result ]")
      val msg = handler?.obtainMessage()
      if (result) {
        msg?.arg1 = 1
      } else {
```

```
        msg?.arg1 = 0
      }
      handler?.sendMessage(msg)
} }
```

这里，我们利用 CRUD 操作结果重新连接了原 sendMessage()方法。11.3 节将对应用程序进行适当改进，并可监听启用、关闭广播消息。

11.3　启用和监听广播

某些时候，在应用程序启动时运行服务是非常重要的；而有时，还需要在终止前执行相应的清理操作。在下列示例中，将扩展 Journaler 应用程序，并对广播消息进行监听，同时执行相关操作。首先，可定义扩展 BroadcastReceiver 类的两个类，如下所示。

- BootReceiver：该类负责处理系统的启动事件。
- ShutdownReceiver：该类负责处理系统的关闭事件。

随后，将其注册至 manifest 文件中，如下所示。

```
<manifest
  ...
>
...
<receiver
    android:name=".receiver.BootReceiver"
    android:enabled="true"
    android:exported="false">
    <intent-filter>
      <action android:name=
      "android.intent.action.BOOT_COMPLETED" />
    </intent-filter>
    <intent-filter>
      <action android:name=
      "android.intent.action.PACKAGE_REPLACED" />
      data android:scheme="package" />
    </intent-filter>
    <intent-filter>
      <action android:name=
      "android.intent.action.PACKAGE_ADDED" />
      <data android:scheme="package" />
    </intent-filter>
```

```
 </receiver>

<receiver android:name=".receiver.ShutdownReceiver">
  <intent-filter>
    <action android:name=
    "android.intent.action.ACTION_SHUTDOWN" />
    <action android:name=
    "android.intent.action.QUICKBOOT_POWEROFF" />
  </intent-filter>
</receiver>
...
</manifest>
```

当启动或替换应用程序时，BootReceiver 类将被触发；而当关闭设备时，关闭操作将被触发。对此，打开 BootReceiver 类，并按照下列方式进行定义：

```
package com.journaler.receiver

import android.content.BroadcastReceiver
import android.content.Context
import android.content.Intent
import android.util.Log

class BootReceiver : BroadcastReceiver() {

  val tag = "Boot receiver"

  override fun onReceive(p0: Context?, p1: Intent?) {
    Log.i(tag, "Boot completed.")
    // Perform your on boot stuff here.
  }

}
```

其中，我们针对上述两个类定义了 receiver 数据包。对于 ShutdownReceiver 类，其定义方式如下所示。

```
package com.journaler.receiver

import android.content.BroadcastReceiver
import android.content.Context
import android.content.Intent
import android.util.Log
```

```
class ShutdownReceiver : BroadcastReceiver() {

  val tag = "Shutdown receiver"

  override fun onReceive(p0: Context?, p1: Intent?) {
    Log.i(tag, "Shutting down.")
    // Perform your on cleanup stuff here.
} }
```

为了保持正常工作，需要对应用程序进行适当的调整，以避免应用程序崩溃。这里，可将 main 服务从 Application 类移至主活动的 onCreate()方法中，这也是 Journaler 类的第一处更新，如下所示。

```
class Journaler : Application() {
  ...
  override fun onCreate() { // We removed start service method
    execution.
    super.onCreate()
    ctx = applicationContext
    Log.v(tag, "[ ON CREATE ]")
  }
  // We removed startService() method implementation.
  ...
}
```

随后，通过在 onCreate()方法的末尾，附加代码行来扩展 MainActivity 类，如下所示。

```
class MainActivity : BaseActivity() {
  ...
  override fun onCreate(savedInstanceState: Bundle?) {
    ...
    val serviceIntent = Intent(this, MainService::class.java)
    startService(serviceIntent)
}
... } }
```

构建并运行应用程序。接下来，关闭并再次打开电话设备，过滤 Logcat 以便仅显示与当前应用程序相关的日志消息。对应输出结果如下所示。

```
... I/Shutdown receiver: Shutting down.
... I/Boot receiver: Boot completed.
```

提示：

某些时候，接收启动事件可能需要些许时间。

11.4　监听网络事件

应用程序的最后一项改进措施是构建连接时执行同步操作。针对于此，可创建在同一 NetworkReceiver 数据包中调用的一个新类，对应实现如下所示。

```
class NetworkReceiver : BroadcastReceiver() {
 private val tag = "Network receiver"
 private var service: MainService? = null
 private val serviceConnection = object : ServiceConnection {
   override fun onServiceDisconnected(p0: ComponentName?) {
     service = null
   }

   override fun onServiceConnected(p0: ComponentName?, binder:
   IBinder?) {
       if (binder is MainService.MainServiceBinder) {
           service = binder.getService()
           service?.synchronize()
       }
   }
  }

  override fun onReceive(context: Context?, p1: Intent?) {
  context?.let {

       val cm = context.getSystemService
       (Context.CONNECTIVITY_SERVICE) as ConnectivityManager

       val activeNetwork = cm.activeNetworkInfo
       val isConnected = activeNetwork != null &&
       activeNetwork.isConnectedOrConnecting
       if (isConnected) {
           Log.v(tag, "Connectivity [ AVAILABLE ]")
           if (service == null) {
               val intent = Intent(context,
               MainService::class.java)
               context.bindService(
                   intent, serviceConnection,
                   android.content.Context.BIND_AUTO_CREATE
               )
           } else {
```

```
                service?.synchronize()
            }
        } else {
            Log.w(tag, "Connectivity [ UNAVAILABLE ]")
            context.unbindService(serviceConnection)
        }
    }
  }
}
```

当产生连接事件时，接收器将接收消息。每次出现一个上下文和连接时，即会绑定到对应服务上并触发同步操作。这里不需要执行频繁的同步触发，在第 12 章中，将在同步方法实现中对此加以保护。下面将更新 Journaler 应用程序类，并注册监听器，如下所示。

```
class Journaler : Application() {
  ...
  override fun onCreate() {
      super.onCreate()
      ctx = applicationContext
      Log.v(tag, "[ ON CREATE ]")
      val filter =
      IntentFilter(ConnectivityManager.CONNECTIVITY_ACTION)
      registerReceiver(networkReceiver, filter)
  }
  ...
}
```

构建并运行应用程序。关闭连接并再次开启连接，Logcat 中的输出结果如下所示。

```
... V/Network receiver: Connectivity [ AVAILABLE ]
... V/Network receiver: Connectivity [ AVAILABLE ]
... V/Network receiver: Connectivity [ AVAILABLE ]
... W/Network receiver: Connectivity [ UNAVAILABLE ]
... V/Network receiver: Connectivity [ AVAILABLE ]
... V/Network receiver: Connectivity [ AVAILABLE ]
... V/Network receiver: Connectivity [ AVAILABLE ]
```

11.5　本 章 小 结

本章探讨了广播消息的应用方式，同时还学习了如何监听系统广播消息，以及自己创建的消息。在此基础上，Journaler 应用程序得到了较大的改进且更具灵活性。第 12 章将通过 Android Framework 进一步丰富应用程序中的内容。

第 12 章　后端和 API

本章将把应用程序连接至远程后端实例上，且全部数据均与后端同步。对于 API 调用，我们将使用 Retrofit。Retrofit 是 Android 平台常用的 HTTP 客户端。本章将通过实践操作方式逐步讲解如何实现与应用程序中后端的连接。

本章篇幅较长，主要涉及以下主题：

- 与数据类协同工作。
- Retrofit。
- 基于 Kotson 库的 Gson。
- 内容供应商。
- 内容加载器。
- Android 适配器。
- 数据绑定。
- 使用列表和网格。

这里，也建议读者认真阅读本章内容，并对应用程序进行不断的尝试。

12.1　确定所用的实体

在同步数据之前，需要确定同步的内容。该问题的答案很明显，但我们依然需要对实体列表进行概括，当前，计划同步两个主要实体：

- Note 实体。
- Todo 实体。

这两个实体包含以下属性：

- 公共属性如下所示。
  - title: String。
  - message: String。
  - location: Location（将被序列化）。

注意：

当前采用数据库中的经纬度表示地理位置。由于针对序列化/反序列化将分别引入

Gson 和 Kotson，因而稍后会将此修改为 Text 类型。

- Todo 属性如下所示。

scheduledFor: Long。

随后，可打开相关类以对上述属性进行查看。

12.2　与数据类协同工作

在 Kotlin 中，建议采用 data 类作为实体的表达方式。在当前示例中，由于扩展了一个公共类，其中包含了 Note 类和 Todo 类之间共享的属性，因而并未使用 data 类。

由于 data 类可简化工作流程，特别是将实体用作后端通信时，因而推荐使用 data 类。

这里，使用类的主要目的是加载数据，同时可自动提供某些功能。下列代码定义了 data 类：

```
data class Entity(val param1: String, val param2: String)
```

对于 data 类，编译器将自动提供下列内容：

- equals()方法和 hashCode()方法。
- 具有可读格式（Entity(param1=Something, param2=Something)）的 toString()方法。
- 用于克隆操作的 copy()方法。

另外，所有 data 类须满足以下条件：

- 主构造方法需要至少包含一个参数。
- 全部的主构造方法参数都需要标记为 val 或 var。
- 数据类不可以为 abstract、open、sealed 或 inner。

接下来将引入某些 data 类。鉴于将使用一个远程后端实例，因而需要获取授权。针对在授权处理过程中传递的数据以及授权结果，我们将定义新的实体（data 类）。对此，创建一个名为 api 的新数据包，随后定义一个名为 UserLoginRequest 的新 data 类，如下所示。

```
package com.journaler.api

data class UserLoginRequest(
  val username: String,
  val password: String
)
```

UserLoginRequest 类中包含了授权证书。API 调用将返回一个 JSON，并被反序列化至 JournalerApiToken 数据类中，如下所示。

```
package com.journaler.api
import com.google.gson.annotations.SerializedName

data class JournalerApiToken(
    @SerializedName("id_token") val token: String,
    val expires: Long
)
```

注意，这里使用了注解通知 Gson，token 字段将从 JSON 的 id_token 字段中获取。

提示：

优先考虑 data 类的使用，特别是数据用于加载数据库和后端信息时。

12.3　将数据模型连接至数据库

假设需要像 Journaler 应用程序那样保存数据库中的数据，并利用远程后端实例进行同步，一种较好的方法是首先构建一个持久化层以存储数据，将数据持久化至本地文件系统数据库中可防止数据丢失，特别是数据量较大时。

我们制定了一个持久化机制，并将全部数据存储至 SQLite 数据库中；随后，本章将引入后端通信机制。由于并不知道 API 调用是否会失败，或者后端实例是否有效，因而需要对数据进行持久化。如果数据仅保存在设备的内存中，如果针对同步操作的 API 调用失败且应用程序崩溃，则数据将丢失；如果数据已被持久化，则可再次尝试同步操作，数据依然存在。

12.4　Retrofit 简介

如前所述，Retrofit 是一个开源库，同时也是较为常见的 Android HTTP 客户端。本节将讨论 Retrofit 方面的知识及其使用方式，对应的版本为 2.3.0。

Retrofit 依赖于其他库，此处将与 Okhttp 结合使用。Okhttp 是一个 HTTP/HTTP2，并由 Retrofit 开发人员发布。接下来首先在 build.gradle 配置中设置依赖关系，如下所示。

```
apply plugin: "com.android.application"
apply plugin: "kotlin-android"
apply plugin: "kotlin-android-extensions"
```

```
...
dependencies {
  ...
  compile 'com.squareup.retrofit2:retrofit:2.3.0'
  compile 'com.squareup.retrofit2:converter-gson:2.0.2'
  compile 'com.squareup.okhttp3:okhttp:3.9.0'
  compile 'com.squareup.okhttp3:logging-interceptor:3.9.0'
}
```

在本书编写时，Retrofit 和 Okhttp 已更新至最新版本。下面针对以下内容添加依赖关系：

- Retrofit 库。
- Gson 转换器，用于反序列化 API 响应。
- Okhttp 库。
- Okhttp 的日志解释器，以输出 API 调用的日志信息。

在同步了 Gradle 配置后，一切均已准备就绪。

12.4.1　定义 Retrofit 服务

Retrofit 将 HTTP API 转换为 Kotlin 接口。在 API 数据包中创建一个名为 JournalerBackendService 的接口，对应代码如下所示。

```
package com.journaler.api

import com.journaler.model.Note
import com.journaler.model.Todo
import retrofit2.Call
import retrofit2.http.*

interface JournalerBackendService {

  @POST("user/authenticate")
  fun login(
        @HeaderMap headers: Map<String, String>,
        @Body payload: UserLoginRequest
  ): Call<JournalerApiToken>

  @GET("entity/note")
  fun getNotes(
        @HeaderMap headers: Map<String, String>
  ): Call<List<Note>>
```

```
    @GET("entity/todo")
    fun getTodos(
          @HeaderMap headers: Map<String, String>
    ): Call<List<Todo>>

    @PUT("entity/note")
    fun publishNotes(
          @HeaderMap headers: Map<String, String>,
          @Body payload: List<Note>
    ): Call<Unit>

    @PUT("entity/todo")
    fun publishTodos(
          @HeaderMap headers: Map<String, String>,
          @Body payload: List<Todo>
    ): Call<Unit>

    @DELETE("entity/note")
    fun removeNotes(
          @HeaderMap headers: Map<String, String>,
          @Body payload: List<Note>
    ): Call<Unit>

    @DELETE("entity/todo")
    fun removeTodos(
          @HeaderMap headers: Map<String, String>,
          @Body payload: List<Todo>
    ): Call<Unit>

}
```

在上述接口中，定义了执行下列操作的调用列表：

- 用户认证：这将接收请求头以及包含用户证书的 UserLoginRequest 类实例，并将用作调用的有效载荷。执行该调用将返回一个封装的 JournalerApiToken 实例。此处需要使用针对其他调用的令牌，并将其内容置于每个调用头中。
- Note 和 Todo 获取：接收包含认证令牌的请求头。作为调用结果，将获得一个封装的 Note 类或 Todo 类实例列表。
- Note 和 Todo 推送（当向服务器发送新数据时）：接收包含认证令牌的请求头。调用的有效载荷表示为 Note 类或 Todo 类实例列表。这里并不会针对此类调用返回任何重要的数据；更为重要的是，响应代码应具备正确性。

- Note 和 Todo 移除：接收包含认证令牌的请求头。调用的有效负载为从远程后端服务器实例中移除的 Note 类或 Todo 类实例列表。这里并不会针对此类调用返回任何重要的数据；更重要的是，响应代码应具有正确性。

每个方法都包含了一个相应的注解，表示包含路径的 HTTP 方法。此外，还将使用注解标记有效载荷体和头映射。

12.4.2 构建 Retrofit 服务实例

在对服务描述完毕后，接下来将构建一个真实的 Retrofit 实例，用于触发 API 调用。对此，首先需要引入某些附加类，并加载 TokenManager 对象中的最新的令牌实例，如下所示。

```
package com.journaler.api
  object TokenManager {
    var currentToken = JournalerApiToken("", -1)
  }
```

另外，还需要设置一个对象，用于获取名为 BackendServiceHeaderMap 的 API 调用头映射，如下所示。

```
  package com.journaler.api

  object BackendServiceHeaderMap {

    fun obtain(authorization: Boolean = false): Map<String, String> {
      val map = mutableMapOf(
             Pair("Accept", "*/*"),
             Pair("Content-Type", "application/json; charset=UTF-8")
      )
      if (authorization) {
         map["Authorization"] = "Bearer
          ${TokenManager.currentToken.token}"
      }
      return map
  }

}
```

下面考查如何构建一个 Retrofit 实例。对此，创建一个名为 BackendServiceRetrofit

的新对象，如下所示。

```
package com.journaler.api

import okhttp3.OkHttpClient
import okhttp3.logging.HttpLoggingInterceptor
import retrofit2.Retrofit
import retrofit2.converter.gson.GsonConverterFactory
import java.util.concurrent.TimeUnit

object BackendServiceRetrofit {

  fun obtain(
        readTimeoutInSeconds: Long = 1,
        connectTimeoutInSeconds: Long = 1
  ): Retrofit {
    val loggingInterceptor = HttpLoggingInterceptor()
    loggingInterceptor.level= HttpLoggingInterceptor.Level.BODY
    return Retrofit.Builder()
            .baseUrl("http://127.0.0.1")
            .addConverterFactory(GsonConverterFactory.create())
            .client(
                    OkHttpClient
                            .Builder()
                            .addInterceptor(loggingInterceptor)
                            .readTimeout(readTimeoutInSeconds,
                            TimeUnit.SECONDS)
                            .connectTimeout
                            (connectTimeoutInSeconds,
                            TimeUnit.SECONDS)
                            .build()
            )
            .build()
  }

}
```

调用 obtain()方法将返回一个 Retrofit 实例，进而引发 API 调用。这里，我们将后端基本 URL 设置为本地主机，同时创建了一个 Retrofit 实例。除此之外，还传递了一个 Gson 转换器工厂，用于 JSON 反序列化机制。更为重要的是，此处传递了一个即将使用的客户端实例，并创建了一个新的 OkHttp 客户端。

12.5　基于 Kotson 库的 Gson

针对每个 Android 应用程序来说，JSON 序列化和反序列化十分重要，且经常被加以使用。有鉴于此，本节将使用谷歌公司发布的 Gson 库。另外，针对 Gson 还将使用 Kotson 和 Kotlin 绑定。

首先需要针对 build.gradle 配置提供相应的依赖关系，如下所示。

```
apply plugin: "com.android.application"
apply plugin: "kotlin-android"
apply plugin: "kotlin-android-extensions"
...
dependencies {
  ...
  compile 'com.google.code.gson:gson:2.8.0'
  compile 'com.github.salomonbrys.kotson:kotson:2.3.0'
  ...
}
```

随后将更新代码，并在数据库管理中使用基于 Kotson 绑定的 Gson 实现地理位置的序列化/反序列化。下面针对 Db 类稍作修改，如下所示。

```
class DbHelper(dbName: String, val version: Int) :
SQLiteOpenHelper(
  Journaler.ctx, dbName, null, version
) {

  companion object {
     val ID: String = "_id"
     val TABLE_TODOS = "todos"
     val TABLE_NOTES = "notes"
     val COLUMN_TITLE: String = "title"
     val COLUMN_MESSAGE: String = "message"
     val COLUMN_LOCATION: String = "location"
     val COLUMN_SCHEDULED: String = "scheduled"
  }
  ...
  private val createTableNotes = """
                              CREATE TABLE if not exists
                              $TABLE_NOTES
                              (
```

```
                                    $ID integer PRIMARY KEY
                                    autoincrement,
                                    $COLUMN_TITLE text,
                                    $COLUMN_MESSAGE text,
                                    $COLUMN_LOCATION text
                                )
                                """

 private val createTableTodos = """
                                CREATE TABLE if not exists
                                $TABLE_TODOS
                                (
                                    $ID integer PRIMARY KEY
                                    autoincrement,
                                    $COLUMN_TITLE text,
                                    $COLUMN_MESSAGE text,
                                    $COLUMN_SCHEDULED integer,
                                    $COLUMN_LOCATION text
                                )
                                """
...
}
```

可以看到，此处修改了地理位置信息处理机制，且未使用地理位置的经纬度列，而是设置了单一数据库列——location，对应的类型为 Text。我们将加载 Gson 库生成的序列化 Location 类值。另外，当检索序列化值时，还将通过 Gson 将其反序列化至 Location 类实例中。

打开 Db.kt，并通过 Gson 对其进行更新，以序列化/反序列化 Locaton 类实例，如下所示。

```
package com.journaler.database
...
import com.google.gson.Gson
...
import com.github.salomonbrys.kotson.*

object Db : Crud<DbModel> {
  ...
  private val gson = Gson()
  ...
  override fun insert(what: Collection<DbModel>): Boolean {
    ...
```

```
what.forEach {
     item ->
     when (item) {
         is Entry -> {
              ...
              values.put(DbHelper.COLUMN_LOCATION,
              gson.toJson(item.location))
              ...
         }
     }
     ...
     return success
}
...
override fun update(what: Collection<DbModel>): Boolean {
    ...
    what.forEach {
         item ->
         when (item) {
             is Entry -> {
                  ...
                  values.put(DbHelper.COLUMN_LOCATION,
                  gson.toJson(item.location))
             }
             ...
             return result
         }
         ...
         override fun select(args: Pair<String, String>, clazz:
         KClass<DbModel>): List<DbModel> {
             return select(listOf(args), clazz)
         }

         override fun select(
             args: Collection<Pair<String, String>>, clazz:Kclass
             <DbModel>
         ): List<DbModel> {
             ...
             if (clazz.simpleName == Note::class.simpleName) {
               val result = mutableListOf<DbModel>()
               val cursor = db.query(
                 ...
               )
```

```
            while (cursor.moveToNext()) {
              ...
              val locationIdx =
              cursor.getColumnIndexOrThrow(DbHelper.COLUMN_LOCATION)
              val locationJson = cursor.getString(locationIdx)
              val location = gson.fromJson<Location>(locationJson)
              val note = Note(title, message, location)
              note.id = id
              result.add(note)
            }
            cursor.close()
            return result
          }
          if (clazz.simpleName == Todo::class.simpleName) {
            ...
          }
          while (cursor.moveToNext()) {
            ...
            val locationIdx =
            cursor.getColumnIndexOrThrow(DbHelper.COLUMN_LOCATION)
            val locationJson = cursor.getString(locationIdx)
            val location = gson.fromJson<Location>(locationJson)
            ...
            val todo = Todo(title, message, location, scheduledFor)
            todo.id = id
            result.add(todo)
          }
          cursor.close()
          return result
        }
        db.close()
        throw IllegalArgumentException("Unsupported entry type:
        $clazz")
      }
    }
  }
}
```

不难发现，基于 Gson 的更新操作十分简单，并依赖于从 Gson 类实例访问的以下两个 Gson 库方法：

- ❑ fromJson<T>()。
- ❑ toJson()。

由于 Kotson 和 Kotlin 绑定，因此可以使用 fromJson<T>()方法对序列化的数据使用参数化类型。

12.6 其 他 方 案

本节将简要介绍 Retrofit 和 Gson 的一些替代方案。在大型开源社区中，每天都会涌现出新鲜的事物，读者有权利进行选择，甚至是构建自己的实现方案。

12.6.1 Retrofit 替代方案

正如 Volley 主页所描述的那样，Volley 是一个 HTTP 库，并简化了 Android 应用程序的网络连接过程；更为重要的是，Volley 具有速度优势。Volley 涵盖了如下一些较为关键的特性：

- 网络请求的自动调度。
- 多个并发网络连接。
- 具有标准 HTTP 缓存一致性的透明磁盘和内存响应缓存。
- 支持请求优先级。
- 取消请求 API。
- 易于定制。
- 强大的命令机制。
- 调试和跟踪工具。

读者可访问 Volley 的主页以了解更多内容，对应网址为 https://github.com/google/volley。

12.6.2 Gson 替代方案

Jackson 是底层的 JSON 解释器，类似于 XML 的 Java StAX 解释器。Jackson 提供了下列关键特性：

- 快速、方便。
- 支持扩展注解。
- 流读取和写入。
- 树形模型。
- 支持 JAX-RS。
- 支持二进制内容。

读者可访问 Jackson 的主页以了解更多内容，对应网址为 https://github.com/FasterXML/jackson。

12.7　执行第一个 API 调用

我们已经定义了包含所有 API 调用的 Retrofit 服务，但截至目前，尚未实现任何连接。本节将对此予以尝试，并扩展代码以使用 Retrofit。其中，每个 API 调用可通过同步或异步方式被执行。前述内容曾将 Retrofit 服务基本 URL 设置为本地主机，这意味着，需要使用本地后端实例响应 HTTP 请求。鉴于后端实现超出了本书的讨论范围，读者可尝试创建一个简单的服务并相应于请求。其间，读者可选择任何一种编程语言，例如 Kotlin、Java、Python 和 PHP。

如果读者不打算实现自己的应用程序来处理 HTTP 请求，则可覆盖基本 URL、Note 和 Todo 路径，并使用已准备就绪的后端实例，如下所示。

- 基本 URL：http://static.milosvasic.net/json/journaler。
- 登录 POST：

```
@POST("authenticate")
// @POST("user/authenticate")
fun login(
    ...
): Call<JournalerApiToken>
```

- Notes GET：

```
@GET("notes")
// @GET("entity/note")
fun getNotes(
    ...
): Call<List<Note>>
```

- Todos GET：

```
  @GET("todos")
 // @GET("entity/todo")
fun getTodos(
    ...
): Call<List<Todo>>
```

类似地，我们将定位返回存根 Note 和 Todo 的远程后端实例。对此，打开

JournalerBackendService 接口，并按照下列方式进行扩展：

```
interface JournalerBackendService {
  companion object {
    fun obtain(): JournalerBackendService {
         return BackendServiceRetrofit
                 .obtain()
                 .create(JournalerBackendService::class.java)
    }
  }
  ...
}
```

上述方法利用 Retrofit 生成 JournalerBackendService 实例。据此，将触发全部调用。接下来打开 MainService 类，并查看 synchronize()方法。其中使用了 sleep 模拟基于后端的通信。下面利用真实的后端调用对其予以替换，如下所示。

```
/**
* Authenticates user synchronously,
* then executes async calls for notes and TODOs fetching.
* Pay attention on synchronously triggered call via execute()method.
* Its asynchronous equivalent is: enqueue().
*/
override fun synchronize() {
    executor.execute {
        Log.i(tag, "Synchronizing data [ START ]")
        var headers = BackendServiceHeaderMap.obtain()
        val service = JournalerBackendService.obtain()
        val credentials = UserLoginRequest("username", "password")
        val tokenResponse = service
                .login(headers, credentials)
                .execute()
        if (tokenResponse.isSuccessful) {
            val token = tokenResponse.body()
            token?.let {
                TokenManager.currentToken = token
                headers = BackendServiceHeaderMap.obtain(true)
                fetchNotes(service, headers)
                fetchTodos(service, headers)
            }
        }
        Log.i(tag, "Synchronizing data [ END ]")
    }
```

```
}

/**
* Fetches notes asynchronously.
* Pay attention on enqueue() method
*/
private fun fetchNotes(
        service: JournalerBackendService, headers: Map<String, String>
) {
    service
        .getNotes(headers)
        .enqueue(
        object : Callback<List<Note>> {
          verride fun onResponse(
           call: Call<List<Note>>?, response: Response<List<Note>>?
                    ) {
                        response?.let {
                            if (response.isSuccessful) {
                                val notes = response.body()
                                notes?.let {
                                    Db.insert(notes)
                                }
                            }
                        }
                    }

                    override fun onFailure(call:
                    Call<List<Note>>?, t: Throwable?) {
                        Log.e(tag, "We couldn't fetch notes.")
                    }
                }
           )
}

/**
* Fetches TODOs asynchronously.
* Pay attention on enqueue() method
*/
private fun fetchTodos(
        service: JournalerBackendService, headers: Map<String, String>
) {
    service
```

```
            .getTodos(headers)
            .enqueue(
                    object : Callback<List<Todo>> {
                        override fun onResponse(
                                call: Call<List<Todo>>?, response:
                                    Response<List<Todo>>?
                        ) {
                            response?.let {
                                if (response.isSuccessful) {
                                    val todos = response.body()
                                    todos?.let {
                                        Db.insert(todos)
                                    }
                                }
                            }
                        }
                        override fun onFailure(call:
                        Call<List<Todo>>?, t: Throwable?) {
                            Log.e(tag, "We couldn't fetch notes.")
                        }
                    }
            )
}
```

在上述代码中，首先创建了头和 Journaler 后端服务实例，随后通过触发 execute()方法执行了同步认证，其中接收了 Response<JournalerApiToken>。JournalerApiToken 实例则封装于 Response 类实例中。在对响应的成功性，以及实际接收并反序列化的 JournalerApiToken 进行检测后，可将其设置为 TokenManager。最后，还将触发对 Note 和 Todo 检索的异步调用。

enqueue()方法将触发异步操作，并作为参数接收 Retrofit 回调。此处执行了与同步调用相同的操作，并对成功性和数据进行检测。如果一切顺利，将把全部实例传递至数据库管理器以用于存储。

这里仅实现了 Note 和 Todo 检索，API 调用的其余部分则留与读者以作练习。这也将是学习 Retrofit 的绝佳方法。

接下来将构建、运行应用程序。随着应用程序及其 main 服务的启动，API 调用也将随之被执行。通过 OkHttp 过滤 Logcat 输出结果，具体如下所示。

认证日志内容如下所示。

- 请求：

```
D/OkHttp: --> POST
http://static.milosvasic.net/jsons/journaler/authenticate
D/OkHttp: Content-Type: application/json; charset=UTF-8
D/OkHttp: Content-Length: 45
D/OkHttp: Accept: */*
D/OkHttp: {"password":"password","username":"username"}
D/OkHttp: --> END POST (45-byte body)
```

❑ 响应：

```
D/OkHttp: <-- 200 OK
http://static.milosvasic.net/jsons/journaler/
authenticate/ (302ms)
D/OkHttp: Date: Sat, 23 Sep 2017 15:46:27 GMT
D/OkHttp: Server: Apache
D/OkHttp: Keep-Alive: timeout=5, max=99
D/OkHttp: Connection: Keep-Alive
D/OkHttp: Transfer-Encoding: chunked
D/OkHttp: Content-Type: text/html
D/OkHttp: {
D/OkHttp: "id_token": "stub_token_1234567",
D/OkHttp: "expires": 10000
D/OkHttp: }
D/OkHttp: <-- END HTTP (58-byte body)
```

Note 日志内容如下所示。

❑ 请求：

```
D/OkHttp: --> GET
http://static.milosvasic.net/jsons/journaler/notes
D/OkHttp: Accept: */*
D/OkHttp: Authorization: Bearer stub_token_1234567
D/OkHttp: --> END GET
```

❑ 响应：

```
D/OkHttp: <-- 200 OK
http://static.milosvasic.net/jsons/journaler/notes/ (95ms)
D/OkHttp: Date: Sat, 23 Sep 2017 15:46:28 GMT
D/OkHttp: Server: Apache
D/OkHttp: Keep-Alive: timeout=5, max=97
D/OkHttp: Connection: Keep-Alive
D/OkHttp: Transfer-Encoding: chunked
```

```
D/OkHttp: Content-Type: text/html
D/OkHttp: [
D/OkHttp: {
D/OkHttp: "title": "Test note 1",
D/OkHttp: "message": "Test message 1",
D/OkHttp: "location": {
D/OkHttp: "latitude": 10000,
D/OkHttp: "longitude": 10000
D/OkHttp: }
D/OkHttp: },
D/OkHttp: {
D/OkHttp: "title": "Test note 2",
D/OkHttp: "message": "Test message 2",
D/OkHttp: "location": {
D/OkHttp: "latitude": 10000,
D/OkHttp: "longitude": 10000
D/OkHttp: }
D/OkHttp: },
D/OkHttp: {
D/OkHttp: "title": "Test note 3",
D/OkHttp: "message": "Test message 3",
D/OkHttp: "location": {
D/OkHttp: "latitude": 10000,
D/OkHttp: "longitude": 10000
D/OkHttp: }
D/OkHttp: }
D/OkHttp: ]
D/OkHttp: <-- END HTTP (434-byte body)
```

Todo 日志内容如下所示。

- 请求示例：

```
D/OkHttp: --> GET
http://static.milosvasic.net/jsons/journaler/todos
D/OkHttp: Accept: */*
D/OkHttp: Authorization: Bearer stub_token_1234567
D/OkHttp: --> END GET
```

- 响应示例：

```
D/OkHttp: <-- 200 OK
http://static.milosvasic.net/jsons/journaler/todos/ (140ms)
D/OkHttp: Date: Sat, 23 Sep 2017 15:46:28 GMT
```

```
D/OkHttp: Server: Apache
D/OkHttp: Keep-Alive: timeout=5, max=99
D/OkHttp: Connection: Keep-Alive
D/OkHttp: Transfer-Encoding: chunked
D/OkHttp: Content-Type: text/html
D/OkHttp: [
D/OkHttp: {
D/OkHttp: "title": "Test todo 1",
D/OkHttp: "message": "Test message 1",
D/OkHttp: "location": {
D/OkHttp: "latitude": 10000,
D/OkHttp: "longitude": 10000
D/OkHttp: },
D/OkHttp: "scheduledFor": 10000
D/OkHttp: },
D/OkHttp: {
D/OkHttp: "title": "Test todo 2",
D/OkHttp: "message": "Test message 2",
D/OkHttp: "location": {
D/OkHttp: "latitude": 10000,
D/OkHttp: "longitude": 10000
D/OkHttp: },
D/OkHttp: "scheduledFor": 10000
D/OkHttp: },
D/OkHttp: {
D/OkHttp: "title": "Test todo 3",
D/OkHttp: "message": "Test message 3",
D/OkHttp: "location": {
D/OkHttp: "latitude": 10000,
D/OkHttp: "longitude": 10000
D/OkHttp: },
D/OkHttp: "scheduledFor": 10000
D/OkHttp: }
D/OkHttp: ]
D/OkHttp: <-- END HTTP (515-byte body)
```

至此，我们实现了第一个服务，接下来将实现其他调用。具体来说，这里将更新服务，以便可接受登录凭证。当前代码对用户名和密码实现了硬编码，对此，可重构代码并传递参数化凭证。

另外，还可对代码进一步改进，以避免在同一时刻多次执行相同的调用，这也是之前工作中遗留的问题。

12.8　内容供应商

本节将进一步改进前述应用程序，同时引入 Android 内容供应商。内容供应商是 Android Framework 须提供的强大特性之一。那么，内容供应商的目的又是什么？顾名思义，内容供应商旨在对存储于应用程序中的数据进行管理，同时提供了与其他应用程序间的数据共享机制，以及数据访问的安全机制。

图 12.1 显示了内容供应商如何对共享存储访问进行管理。

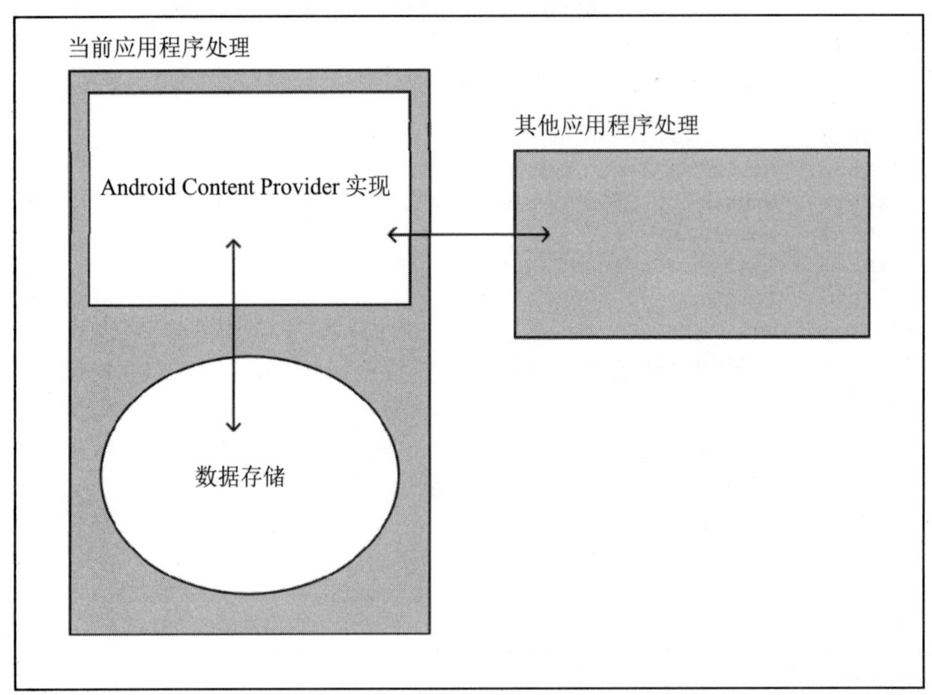

图 12.1

这里，我们计划将 Note 和 Todo 与其他应用程序共享。由于内容供应商提供了抽象层，因而可方便地在存储实现层中进行修改，且不会对上方各层带来影响。因此，即使不打算与其他应用程序共享任何数据，也可以使用内容供应商。例如，可以将 SQLite 中的持久性机制替换为完全不同的机制，如图 12.2 所示。

如果读者不确定是否需要内容供应商，可参考以下几点建议：

- 计划与其他应用程序共享当前应用程序的数据。

- ❑ 在应用程序间复制和粘贴复杂的数据或文件。
- ❑ 支持自定义搜索建议。

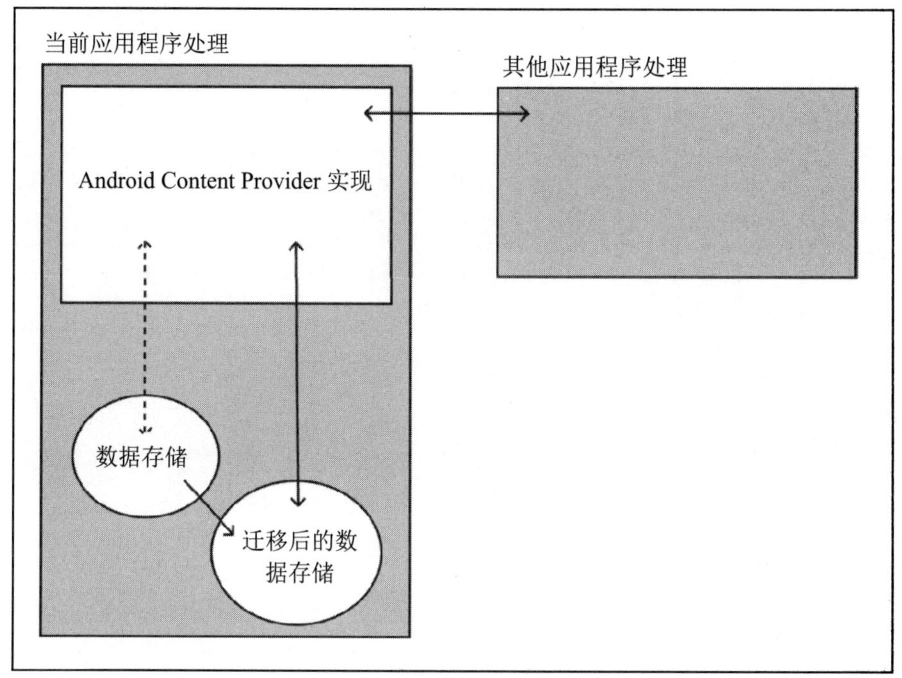

图 12.2

Android 框架提供了一个已经定义好的、可供使用的内容供应商，如管理联系人、音频、视频或其他文件。内容供应商不仅限于 SQLite 访问，还可针对其他结构化数据加以使用。

内容供应商的优点总结如下：

- ❑ 访问数据的权限。
- ❑ 抽象数据层。

如前所述，我们计划公开来自 Journaler 应用程序的数据。在创建内容供应商之前，需要对当前代码进行重构。随后，将创建一个示例客户端应用程序，并使用内容供应商触发所有的 CRUD 操作。

下面定义 ContentProvider 类。对此，创建一个名为 provider 的新数据包，并定义扩展 ContentProvider 类的 JournalerProvider 类。

类定义如下所示。

```
package com.journaler.provider

import android.content.*
import android.database.Cursor
import android.net.Uri
import com.journaler.database.DbHelper
import android.content.ContentUris
import android.database.SQLException
import android.database.sqlite.SQLiteDatabase
import android.database.sqlite.SQLiteQueryBuilder
import android.text.TextUtils

class JournalerProvider : ContentProvider() {

  private val version = 1
  private val name = "journaler"
  private val db: SQLiteDatabase by lazy {
    DbHelper(name, version).writableDatabase
}
```

定义 companion 对象，如下所示。

```
companion object {
    private val dataTypeNote = "note"
    private val dataTypeNotes = "notes"
    private val dataTypeTodo = "todo"
    private val dataTypeTodos = "todos"
    val AUTHORITY = "com.journaler.provider"
    val URL_NOTE = "content://$AUTHORITY/$dataTypeNote"
    val URL_TODO = "content://$AUTHORITY/$dataTypeTodo"
    val URL_NOTES = "content://$AUTHORITY/$dataTypeNotes"
    val URL_TODOS = "content://$AUTHORITY/$dataTypeTodos"
    private val matcher = UriMatcher(UriMatcher.NO_MATCH)
    private val NOTE_ALL = 1
    private val NOTE_ITEM = 2
    private val TODO_ALL = 3
    private val TODO_ITEM = 4
}
```

类实例化，如下所示。

```
/**
* We register uri paths in the following format:
*
```

```
* <prefix>://<authority>/<data_type>/<id>
* <prefix> - This is always set to content://
* <authority> - Name for the content provider
* <data_type> - The type of data we provide in this Uri
* <id> - Record ID.
*/
init {
   /**
    * The calls to addURI() go here,
    * for all of the content URI patterns that the provider should
    * recognize.
    *
    * First:
    *
    * Sets the integer value for multiple rows in notes (TODOs) to 1.
    * Notice that no wildcard is used in the path.
    *
    * Second:
    *
    * Sets the code for a single row to 2. In this case, the "#"
    * wildcard is used.
    * "content://com.journaler.provider/note/3" matches, but
    * "content://com.journaler.provider/note doesn't.
    *
    * The same applies for TODOs.
    *
    * addUri() params:
    *
    * authority - String: the authority to match
    *
    * path - String: the path to match.
    * * may be used as a wild card for any text,
    * and # may be used as a wild card for numbers.
    *
    * code - int: the code that is returned when a URI
    * is matched against the given components.
    */
   matcher.addURI(AUTHORITY, dataTypeNote, NOTE_ALL)
   matcher.addURI(AUTHORITY, "$dataTypeNotes/#", NOTE_ITEM)
   matcher.addURI(AUTHORITY, dataTypeTodo, TODO_ALL)
   matcher.addURI(AUTHORITY, "$dataTypeTodos/#", TODO_ITEM)
}
```

重载 onCreate()方法，如下所示。

```
/**
 * True - if the provider was successfully loaded
 */
override fun onCreate() = true
```

插入操作如下所示。

```
override fun insert(uri: Uri?, values: ContentValues?): Uri {
    uri?.let {
        values?.let {
            db.beginTransaction()
            val (url, table) = getParameters(uri)
            if (!TextUtils.isEmpty(table)) {
                val inserted = db.insert(table, null, values)
                val success = inserted > 0
                if (success) {
                    db.setTransactionSuccessful()
                }
                db.endTransaction()
                if (success) {
                    val resultUrl = ContentUris.withAppendedId
                    (Uri.parse(url), inserted)
                    context.contentResolver.notifyChange(resultUrl,
                    null)
                    return resultUrl
                }
            } else {
                throw SQLException("Insert failed, no table for
                uri: " + uri)
            }
        }
    }
    throw SQLException("Insert failed: " + uri)
}
```

更新操作如下所示。

```
override fun update(
        uri: Uri?,
        values: ContentValues?,
        where: String?,
        whereArgs: Array<out String>?
```

```
): Int {
    uri?.let {
        values?.let {
            db.beginTransaction()
            val (_, table) = getParameters(uri)
            if (!TextUtils.isEmpty(table)) {
                val updated = db.update(table, values, where,
                 whereArgs)
                val success = updated > 0
                if (success) {
                    db.setTransactionSuccessful()
                }
                db.endTransaction()
                if (success) {
                    context.contentResolver.notifyChange(uri, null)
                    return updated
                }
            } else {
                throw SQLException("Update failed, no table for
                 uri: " + uri)
            }
        }
    }
    throw SQLException("Update failed: " + uri)
}
```

删除操作如下所示。

```
override fun delete(
        uri: Uri?,
        selection: String?,
        selectionArgs: Array<out String>?
): Int {
    uri?.let {
        db.beginTransaction()
        val (_, table) = getParameters(uri)
        if (!TextUtils.isEmpty(table)) {
            val count = db.delete(table, selection, selectionArgs)
            val success = count > 0
            if (success) {
                db.setTransactionSuccessful()
            }
            db.endTransaction()
            if (success) {
```

```
                context.contentResolver.notifyChange(uri, null)
                return count
            }
        } else {
            throw SQLException("Delete failed, no table for uri: "
            + uri)
        }
    }
    throw SQLException("Delete failed: " + uri)
}
```

查询操作如下所示。

```
override fun query(
        uri: Uri?,
        projection: Array<out String>?,
        selection: String?,
        selectionArgs: Array<out String>?,
        sortOrder: String?
): Cursor {
    uri?.let {
        val stb = SQLiteQueryBuilder()
        val (_, table) = getParameters(uri)
        stb.tables = table
        stb.setProjectionMap(mutableMapOf<String, String>())
        val cursor = stb.query(db, projection, selection,
          selectionArgs, null, null, null)
        // register to watch a content URI for changes
        cursor.setNotificationUri(context.contentResolver, uri)
        return cursor
    }
    throw SQLException("Query failed: " + uri)
}

/**
* Return the MIME type corresponding to a content URI.
*/
override fun getType(p0: Uri?): String = when (matcher.match(p0)) {
    NOTE_ALL -> {
        "${ContentResolver.
        CURSOR_DIR_BASE_TYPE}/vnd.com.journaler.note.items"
    }
    NOTE_ITEM -> {
```

```
        "${ContentResolver.
        CURSOR_ITEM_BASE_TYPE}/vnd.com.journaler.note.item"
    }
    TODO_ALL -> {
        "${ContentResolver.
        CURSOR_DIR_BASE_TYPE}/vnd.com.journaler.todo.items"
    }
    TODO_ITEM -> {
        "${ContentResolver.
        CURSOR_ITEM_BASE_TYPE}/vnd.com.journaler.todo.item"
    }
    else -> throw IllegalArgumentException
    ("Unsupported Uri [ $p0 ]")
}
```

类的结尾部分如下所示。

```
private fun getParameters(uri: Uri): Pair<String, String> {
    if (uri.toString().startsWith(URL_NOTE)) {
        return Pair(URL_NOTE, DbHelper.TABLE_NOTES)
    }
    if (uri.toString().startsWith(URL_NOTES)) {
        return Pair(URL_NOTES, DbHelper.TABLE_NOTES)
    }
    if (uri.toString().startsWith(URL_TODO)) {
        return Pair(URL_TODO, DbHelper.TABLE_TODOS)
    }
    if (uri.toString().startsWith(URL_TODOS)) {
        return Pair(URL_TODOS, DbHelper.TABLE_TODOS)
    }
    return Pair("", "")
  }

}
```

上述代码实现了以下任务：

- 定义了数据库名称和版本。
- 定义了数据库实例延迟初始化。
- 定义了访问数据的 URI。
- 实现了全部 CRUD 操作。
- 定义了数据的 MIME 类型。

对于内容供应商实现，还需要将其注册于 manifest 中，如下所示。

```
<manifest xmlns:android=
"http://schemas.android.com/apk/res/android"
package="com.journaler">
...
  <application
    ...
  >
    ...
    <provider
         android:exported="true"
         android:name="com.journaler.provider.JournalerProvider"
         android:authorities="com.journaler.provider" />
    ...
  </application>
...
</manifest>
```

这里将 exported 属性设置为 true。这意味着，若为 true，则 Journaler 供应商对于其他应用程序有效。任何应用程序均可使用该供应商的内容 URI 访问数据。另一个较为重要的属性是 multiprocess。如果应用程序运行于多个线程中，该属性负责确定是否创建 Journaler 供应商的多个实例。若为 true，则每个应用程序进程均包含自身的内容供应商实例。

在 Crud 接口中，将此添加至 companion 对象中（如果不存在），如下所示。

```
companion object {
   val BROADCAST_ACTION = "com.journaler.broadcast.crud"
   val BROADCAST_EXTRAS_KEY_CRUD_OPERATION_RESULT = "crud_result"
}
```

这里，将 Db 类重命名为 Content。下面更新 Content 实现，并使用 JournalerProvider，如下所示。

```
package com.journaler.database

import android.content.ContentValues
import android.location.Location
import android.net.Uri
import android.util.Log
import com.github.salomonbrys.kotson.fromJson
import com.google.gson.Gson
import com.journaler.Journaler
import com.journaler.model.*
import com.journaler.provider.JournalerProvider
```

```
object Content {

  private val gson = Gson()
  private val tag = "Content"

  val NOTE = object : Crud<Note> { ...
```

Note 的插入操作如下所示。

```
...
override fun insert(what: Note): Long {
  val inserted = insert(listOf(what))
  if (!inserted.isEmpty()) return inserted[0]
     return 0
}

override fun insert(what: Collection<Note>): List<Long> {
    val ids = mutableListOf<Long>()
    what.forEach { item ->
       val values = ContentValues()
       values.put(DbHelper.COLUMN_TITLE, item.title)
       values.put(DbHelper.COLUMN_MESSAGE, item.message)
       values.put(DbHelper.COLUMN_LOCATION,
       gson.toJson(item.location))
       val uri = Uri.parse(JournalerProvider.URL_NOTE)
       val ctx = Journaler.ctx
       ctx?.let {
         val result = ctx.contentResolver.insert(uri, values)
         result?.let {
              try {
                  ids.add(result.lastPathSegment.toLong())
              } catch (e: Exception) {
                 Log.e(tag, "Error: $e")
              }
         }
       }
    }
    return ids
} ...
```

Note 的更新操作如下所示。

```
...
```

```
override fun update(what: Note) = update(listOf(what))

override fun update(what: Collection<Note>): Int {
  var count = 0
  what.forEach { item ->
       val values = ContentValues()
       values.put(DbHelper.COLUMN_TITLE, item.title)
       values.put(DbHelper.COLUMN_MESSAGE, item.message)
       values.put(DbHelper.COLUMN_LOCATION,
       gson.toJson(item.location))
       val uri = Uri.parse(JournalerProvider.URL_NOTE)
       val ctx = Journaler.ctx
       ctx?.let {
         count += ctx.contentResolver.update(
           uri, values, "_id = ?", arrayOf(item.id.toString())
         )
       }
  }
  return count
} ...
```

Note 的删除操作如下所示。

```
...
override fun delete(what: Note): Int = delete(listOf(what))

  override fun delete(what: Collection<Note>): Int {
    var count = 0
    what.forEach { item ->
      val uri = Uri.parse(JournalerProvider.URL_NOTE)
      val ctx = Journaler.ctx
      ctx?.let {
        count += ctx.contentResolver.delete(
        uri, "_id = ?", arrayOf(item.id.toString())
      )
    }
  }
  return count
} ...
```

Note 的选择操作如下所示。

```
...
override fun select(args: Pair<String, String>
```

```
): List<Note> = select(listOf(args))

override fun select(args: Collection<Pair<String, String>>):
List<Note> {
      val items = mutableListOf<Note>()
      val selection = StringBuilder()
      val selectionArgs = mutableListOf<String>()
      args.forEach { arg ->
          selection.append("${arg.first} == ?")
          selectionArgs.add(arg.second)
      }
      val ctx = Journaler.ctx
      ctx?.let {
          val uri = Uri.parse(JournalerProvider.URL_NOTES)
          val cursor = ctx.contentResolver.query(
                  uri, null, selection.toString(),
            selectionArgs.toTypedArray(), null
          )
          while (cursor.moveToNext()) {
              val id = cursor.getLong
              (cursor.getColumnIndexOrThrow(DbHelper.ID))
              val titleIdx = cursor.getColumnIndexOrThrow
              (DbHelper.COLUMN_TITLE)
              val title = cursor.getString(titleIdx)
              val messageIdx = cursor.getColumnIndexOrThrow
              (DbHelper.COLUMN_MESSAGE)
              val message = cursor.getString(messageIdx)
              val locationIdx = cursor.getColumnIndexOrThrow
              (DbHelper.COLUMN_LOCATION)
              val locationJson = cursor.getString(locationIdx)
              val location = gson.fromJson<Location>
              (locationJson)
              val note = Note(title, message, location)
              note.id = id
              items.add(note)
          }
          cursor.close()
          return items
      }
      return items
}
```

```
override fun selectAll(): List<Note> {
    val items = mutableListOf<Note>()
    val ctx = Journaler.ctx
    ctx?.let {
        val uri = Uri.parse(JournalerProvider.URL_NOTES)
        val cursor = ctx.contentResolver.query(
                uri, null, null, null, null
        )
        while (cursor.moveToNext()) {
            val id = cursor.getLong
            (cursor.getColumnIndexOrThrow(DbHelper.ID))
            val titleIdx = cursor.getColumnIndexOrThrow
            (DbHelper.COLUMN_TITLE)
            val title = cursor.getString(titleIdx)
            val messageIdx = cursor.getColumnIndexOrThrow
            (DbHelper.COLUMN_MESSAGE)
            val message = cursor.getString(messageIdx)
            val locationIdx = cursor.getColumnIndexOrThrow
            (DbHelper.COLUMN_LOCATION)
            val locationJson = cursor.getString(locationIdx)
            val location = gson.fromJson<Location>
            (locationJson)
            val note = Note(title, message, location)
            note.id = id
            items.add(note)
        }
        cursor.close()
    }
    return items
  }
}
```

Todo 对象的定义及其插入操作如下所示。

```
...
val TODO = object : Crud<Todo> {
  override fun insert(what: Todo): Long {
      val inserted = insert(listOf(what))
      if (!inserted.isEmpty()) return inserted[0]
      return 0
  }

  override fun insert(what: Collection<Todo>): List<Long> {
```

```
    val ids = mutableListOf<Long>()
    what.forEach { item ->
        val values = ContentValues()
        values.put(DbHelper.COLUMN_TITLE, item.title)
        values.put(DbHelper.COLUMN_MESSAGE, item.message)
        values.put(DbHelper.COLUMN_LOCATION,
        gson.toJson(item.location))
        val uri = Uri.parse(JournalerProvider.URL_TODO)
        values.put(DbHelper.COLUMN_SCHEDULED,
        item.scheduledFor)
        val ctx = Journaler.ctx
        ctx?.let {
            val result = ctx.contentResolver.insert(uri,
            values)
            result?.let {
                try {
                    ids.add(result.lastPathSegment.toLong())
                } catch (e: Exception) {
                    Log.e(tag, "Error: $e")
                }
            }
        }
    }
    return ids
} ...
```

Todo 的更新操作如下所示。

```
...
override fun update(what: Todo) = update(listOf(what))

override fun update(what: Collection<Todo>): Int {
  var count = 0
  what.forEach { item ->
          val values = ContentValues()
          values.put(DbHelper.COLUMN_TITLE, item.title)
          values.put(DbHelper.COLUMN_MESSAGE, item.message)
          values.put(DbHelper.COLUMN_LOCATION,
          gson.toJson(item.location))
          val uri = Uri.parse(JournalerProvider.URL_TODO)
          values.put(DbHelper.COLUMN_SCHEDULED,
          item.scheduledFor)
          val ctx = Journaler.ctx
```

```
            ctx?.let {
                count += ctx.contentResolver.update(
                        uri, values, "_id = ?",
                       arrayOf(item.id.toString())
                )
            }
    }
    return count
} ...
```

Todo 的删除操作如下所示。

```
...
override fun delete(what: Todo): Int = delete(listOf(what))

override fun delete(what: Collection<Todo>): Int {
     var count = 0
     what.forEach { item ->
          val uri = Uri.parse(JournalerProvider.URL_TODO)
          val ctx = Journaler.ctx
          ctx?.let {
              count += ctx.contentResolver.delete(
                      uri, "_id = ?", arrayOf(item.id.toString())
              )
          }
     }
     return count
}
```

Todo 的选择操作如下所示。

```
...
override fun select(args: Pair<String, String>): List<Todo> =
select(listOf(args))

override fun select(args: Collection<Pair<String, String>>):
  List<Todo> {
    val items = mutableListOf<Todo>()
    val selection = StringBuilder()
    val selectionArgs = mutableListOf<String>()
    args.forEach { arg ->
      selection.append("${arg.first} == ?")
      selectionArgs.add(arg.second)
    }
```

```
    val ctx = Journaler.ctx
    ctx?.let {
      val uri = Uri.parse(JournalerProvider.URL_TODOS)
      val cursor = ctx.contentResolver.query(
            uri, null, selection.toString(),
            selectionArgs.toTypedArray(), null
      )
      while (cursor.moveToNext()) {
          val id = cursor.getLong
          (cursor.getColumnIndexOrThrow(DbHelper.ID))
          val titleIdx = cursor.getColumnIndexOrThrow
          (DbHelper.COLUMN_TITLE)
          val
          title =
          cursor.getString(titleIdx)
          val messageIdx = cursor.getColumnIndexOrThrow
          (DbHelper.COLUMN_MESSAGE)
          val message = cursor.getString(messageIdx)
          val locationIdx = cursor.getColumnIndexOrThrow
          (DbHelper.COLUMN_LOCATION)
          val locationJson = cursor.getString(locationIdx)
          val location = gson.fromJson<Location>
          (locationJson)
          val scheduledForIdx = cursor.getColumnIndexOrThrow(
             DbHelper.COLUMN_SCHEDULED
          )
          val scheduledFor = cursor.getLong(scheduledForIdx)
          val todo = Todo(title, message, location,
          scheduledFor)
          todo.id = id
          items.add(todo)
      }
      cursor.close()
    }
    return items
}

override fun selectAll(): List<Todo> {
    val items = mutableListOf<Todo>()
    val ctx = Journaler.ctx
    ctx?.let {
      val uri = Uri.parse(JournalerProvider.URL_TODOS)
```

```
        val cursor = ctx.contentResolver.query(
                uri, null, null, null, null
        )
        while (cursor.moveToNext()) {
            val id = cursor.getLong
            (cursor.getColumnIndexOrThrow(DbHelper.ID))
            val titleIdx = cursor.getColumnIndexOrThrow
            (DbHelper.COLUMN_TITLE)
            val title = cursor.getString(titleIdx)
            val messageIdx = cursor.getColumnIndexOrThrow
            (DbHelper.COLUMN_MESSAGE)
            val message = cursor.getString(messageIdx)
            val locationIdx = cursor.getColumnIndexOrThrow
            (DbHelper.COLUMN_LOCATION)
            val locationJson = cursor.getString(locationIdx)
            val location = gson.fromJson<Location>
            (locationJson)
            val scheduledForIdx = cursor.getColumnIndexOrThrow(
                DbHelper.COLUMN_SCHEDULED
            )
            val scheduledFor = cursor.getLong(scheduledForIdx)
            val todo = Todo
            (title, message, location, scheduledFor)
            todo.id = id
            items.add(todo)
        }
        cursor.close()
      }
      return items
    }
  }
}
```

代码中利用内容供应商替换了直接数据库访问。随后，更新 UI 类，并使用最新重构的代码。此外，读者还可访问 GitHub 以了解更多内容，对应网址为 https:/github.com/PacktPublishing/-Mastering-Android-Development-with-Kotlin/tree/examples/chapter_12。

其中还包含了一个 Journaler 内容供应商客户端应用程序的示例。下面将在客户端应用程序的主屏幕上突出显示一个使用示例，该示例中包含了 4 个按钮，每个按钮触发一个 CRUD 操作示例，如下所示。

```
package com.journaler.content_provider_client
```

```
import android.content.ContentValues
import android.location.Location
import android.net.Uri
import android.os.AsyncTask
import android.os.Bundle
import android.support.v7.app.AppCompatActivity
import android.util.Log
import com.github.salomonbrys.kotson.fromJson
import com.google.gson.Gson
import kotlinx.android.synthetic.main.activity_main.*

class MainActivity : AppCompatActivity() {

  private val gson = Gson()
  private val tag = "Main activity"

  override fun onCreate(savedInstanceState: Bundle?) {
     super.onCreate(savedInstanceState)
     setContentView(R.layout.activity_main)

     select.setOnClickListener {
       val task = object : AsyncTask<Unit, Unit, Unit>() {
          override fun doInBackground(vararg p0: Unit?) {
               val selection = StringBuilder()
               val selectionArgs = mutableListOf<String>()
               val uri = Uri.parse
               ("content://com.journaler.provider/notes")
               val cursor = contentResolver.query(
                       uri, null, selection.toString(),
                       selectionArgs.toTypedArray(), null
               )
               while (cursor.moveToNext()) {
                    val id = cursor.getLong
                    (cursor.getColumnIndexOrThrow("_id"))
                    val titleIdx = cursor.
                    getColumnIndexOrThrow("title")
                    val title = cursor.getString(titleIdx)
                    val messageIdx = cursor.
                    getColumnIndexOrThrow("message")
                    val message = cursor.getString(messageIdx)
                    val locationIdx = cursor.
                    getColumnIndexOrThrow("location")
```

```
                val locationJson = cursor.
                getString(locationIdx)
                val location =
                gson.fromJson<Location>(locationJson)
                Log.v(
                        tag,
                        "Note retrieved via content provider [
                        $id, $title, $message, $location ]"
                )
            }
            cursor.close()
        }
    }
    task.execute()
}

insert.setOnClickListener {
    val task = object : AsyncTask<Unit, Unit, Unit>() {
        override fun doInBackground(vararg p0: Unit?) {
            for (x in 0..5) {
                val uri = Uri.parse
                ("content://com.journaler.provider/note")
                val values = ContentValues()
                values.put("title", "Title $x")
                values.put("message", "Message $x")
                val location = Location("stub location $x")
                location.latitude = x.toDouble()
                location.longitude = x.toDouble()
                values.put("location", gson.toJson(location))
                if (contentResolver.insert(uri, values) !=
                null) {
                    Log.v(
                            tag,
                            "Note inserted [ $x ]"
                    )
                } else {
                    Log.e(
                            tag,
                            "Note not inserted [ $x ]"
                    )
                }
            }
```

```
        }
    }
    task.execute()
}

update.setOnClickListener {
    val task = object : AsyncTask<Unit, Unit, Unit>() {
        override fun doInBackground(vararg p0: Unit?) {
            val selection = StringBuilder()
            val selectionArgs = mutableListOf<String>()
            val uri =
            Uri.parse("content://com.journaler.provider/notes")
            val cursor = contentResolver.query(
                    uri, null, selection.toString(),
                    selectionArgs.toTypedArray(), null
            )
            while (cursor.moveToNext()) {
                val values = ContentValues()
                val id = cursor.getLong
                (cursor.getColumnIndexOrThrow("_id"))
                val titleIdx =
                cursor.getColumnIndexOrThrow("title")
                val title = "${cursor.getString(titleIdx)} upd:
                ${System.currentTimeMillis()}"
                val messageIdx =
                cursor.getColumnIndexOrThrow("message")
                val message =
                "${cursor.getString(messageIdx)} upd:
                ${System.currentTimeMillis()}"
                val locationIdx =
                cursor.getColumnIndexOrThrow("location")
                val locationJson =
                cursor.getString(locationIdx)
                values.put("_id", id)
                values.put("title", title)
                values.put("message", message)
                values.put("location", locationJson)

                val updated = contentResolver.update(
                        uri, values, "_id = ?",
                        arrayOf(id.toString())
                )
```

```
                if (updated > 0) {
                    Log.v(
                            tag,
                            "Notes updated [ $updated ]"
                    )
                } else {
                    Log.e(
                            tag,
                            "Notes not updated"
                    )
                }
            }
            cursor.close()
        }
    }
    task.execute()
}

delete.setOnClickListener {
    val task = object : AsyncTask<Unit, Unit, Unit>() {
        override fun doInBackground(vararg p0: Unit?) {
            val selection = StringBuilder()
            val selectionArgs = mutableListOf<String>()
            val uri = Uri.parse
            ("content://com.journaler.provider/notes")
            val cursor = contentResolver.query(
                    uri, null, selection.toString(),
                    selectionArgs.toTypedArray(), null
            )
            while (cursor.moveToNext()) {
                val id = cursor.getLong
                (cursor.getColumnIndexOrThrow("_id"))
                val deleted = contentResolver.delete(
                        uri, "_id = ?", arrayOf(id.toString())
                )
                if (deleted > 0) {
                    Log.v(
                            tag,
                            "Notes deleted [ $deleted ]"
                    )
                } else {
                    Log.e(
```

```
                                    tag,
                                    "Notes not deleted"
                                )
                            }
                        }
                        cursor.close()
                    }
                }
            task.execute()
        }
    }
}
```

该示例展示了如何利用内容供应商从其他应用程序中触发 CRUD 操作。

12.9　Android 适配器

当显示主屏幕上的内容时，需要使用 Android Adapter 类。作为一种机制，Android Framework 提供了适配器，进而作为列表或网格向视图分组中提供相关条目。当展示 Adapter 应用示例时，需要定义自己的适配器实现。对此，创建一个名为 adapter 的新数据包，以及定义一个扩展 BaseAdapter 类的 EntryAdapter 成员类，如下所示。

```
package com.journaler.adapter

import android.annotation.SuppressLint
import android.content.Context
import android.view.LayoutInflater
import android.view.View
import android.view.ViewGroup
import android.widget.BaseAdapter
import android.widget.TextView
import com.journaler.R
import com.journaler.model.Entry

class EntryAdapter(
  private val ctx: Context,
  private val items: List<Entry>
) : BaseAdapter() {

  @SuppressLint("InflateParams", "ViewHolder")
```

```
override fun getView(p0: Int, p1: View?, p2: ViewGroup?): View {
  p1?.let {
      return p1
  }
  val inflater = LayoutInflater.from(ctx)
  val view = inflater.inflate(R.layout.adapter_entry, null)
  val label = view.findViewById<TextView>(R.id.title)
  label.text = items[p0].title
  return view
}

override fun getItem(p0: Int): Entry = items[p0]
override fun getItemId(p0: Int): Long = items[p0].id
override fun getCount(): Int = items.size
}
```

其中重载了下列方法：

- getView()：该方法根据当前容器位置返回填充后的视图实例。
- getItem()：该方法返回用于创建当前视图的条目实例；在当前示例中，这表示为 Entry 类实例（Note 或 Todo）。
- getItemId()：该方法返回当前条目实例的 ID。
- getCount()：该方法返回全部条目数量。

接下来将连接适配器和 UI。对此，打开 ItemsFragment，更新其 onResume()方法，实例化适配器并将其分配与一个 ListView，如下所示。

```
override fun onResume() {
    super.onResume()
    ...
    executor.execute {
        val notes = Content.NOTE.selectAll()
        val adapter = EntryAdapter(activity, notes)
        activity.runOnUiThread {
            view?.findViewById<ListView>(R.id.items)?.adapter =
         adapter
        }
    }
}
```

当绑定和运行应用程序时，可以看到每个 ViewPager 页面均填充了加载后的条目，如图 12.3 所示。

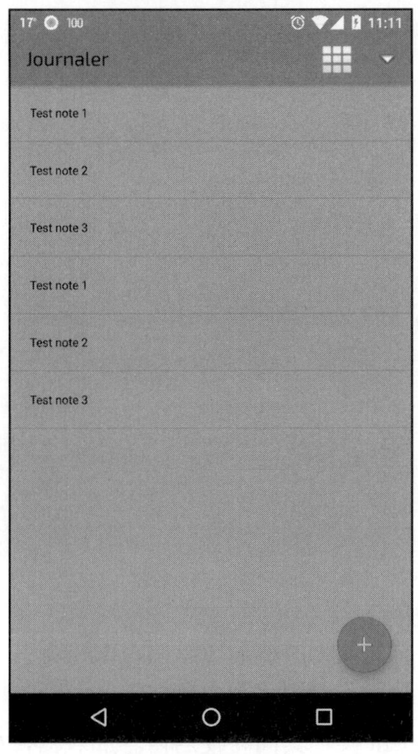

图 12.3

12.10　内容加载器

内容加载器提供了某种机制，可从内容供应商或其他数据源加载数据，进而实现 UI 组件、活动或片段中的显示功能。加载器涵盖了以下优点：

- 运行于独立的线程中。
- 通过回调方法简化了线程的管理。
- 加载器可在配置变化的情况下持久化和缓存结果，进而防止重复查询。
- 可以实现并作为观察者来监视数据中的变化。

接下来将创建内容加载器实现。首先需要更新 Adapter 类。由于要对游标进行处理，因而这里将使用 CursorAdapter 而非 BaseAdapter。相应地，CursorAdapter 接收一个 Cursor 实例作为主构造方法中的参数。CursorAdapter 的实现过程较为简单，对此，可打开 EntryAdapter 并按照下列方式将其进行更新：

```
class EntryAdapter(ctx: Context, crsr: Cursor) : CursorAdapter(ctx,
crsr) {

  override fun newView(p0: Context?, p1: Cursor?, p2: ViewGroup?):
  View {
     val inflater = LayoutInflater.from(p0)
     return inflater.inflate(R.layout.adapter_entry, null)
  }

  override fun bindView(p0: View?, p1: Context?, p2: Cursor?) {
     p0?.let {
          val label = p0.findViewById<TextView>(R.id.title)
          label.text = cursor.getString(
               cursor.getColumnIndexOrThrow(DbHelper.COLUMN_TITLE)
          )
     }
  }
}
```

此处重载了以下两个方法：

- newView()：该方法将返回用数据填充的视图实例。
- bindView()：该方法向 Cursor 实例中填充数据。

最后，更新 ItemsFragment 类，其间将使用内容加载器，如下所示。

```
class ItemsFragment : BaseFragment() {
  ...
  private var adapter: EntryAdapter? = null
  ...
  private val loaderCallback = object :
  LoaderManager.LoaderCallbacks<Cursor> {
    override fun onLoadFinished(loader: Loader<Cursor>?, cursor:
    Cursor?) {
         cursor?.let {
              if (adapter == null) {
                   adapter = EntryAdapter(activity, cursor)
                   items.adapter = adapter
              } else {
                   adapter?.swapCursor(cursor)
              }
         }
    }

    override fun onLoaderReset(loader: Loader<Cursor>?) {
```

```
        adapter?.swapCursor(null)
    }

    override fun onCreateLoader(id: Int, args: Bundle?):
    Loader<Cursor> {
        return CursorLoader(
                activity,
                Uri.parse(JournalerProvider.URL_NOTES),
                null,
                null,
                null,
                null
        )
    }
  }

  override fun onCreate(savedInstanceState: Bundle?) {
     super.onCreate(savedInstanceState)
     loaderManager.initLoader(
             0, null, loaderCallback
     )
  }

  override fun onResume() {
     super.onResume()
     loaderManager.restartLoader(0, null, loaderCallback)
     val btn = view?.findViewById
     <FloatingActionButton>(R.id.new_item)
     btn?.let {
        animate(btn, false)
     }
  }
}
```

这里，通过调用 Fragment 的 LoaderManager 成员初始化了 LoaderManager，并执行了以下两个较为重要的方法：

- initLoader()：该方法确保加载器被初始化并处于活动状态。
- restartLoader()：该方法将启动一个新的 loader 实例，或者重新启动一个现有的 loader 实例。

上述两个方法都接收加载器 ID 和 bundle 数据作为参数，以及 LoaderCallbacks<Cursor> 实现，它提供了以下 3 种重载方法：

- onCreateLoader()：针对所提供的 ID，该方法实例化并返回一个新的加载器实例。
- onLoadFinished()：当之前创建的一个加载器完成了加载时，将调用该方法。
- onLoaderRest()：当之前创建的一个加载器正在重置时，将调用该方法，因此，使其数据无效。

12.11　数 据 绑 定

Android 支持数据绑定机制，以使数据可绑定至视图上，同时最小化黏合代码量。通过更新 Gradle 构建配置，即可启用数据绑定机制，如下所示。

```
android {
  ....
  dataBinding {
    enabled = true
  }
}
...
dependencies {
  ...
  kapt 'com.android.databinding:compiler:2.3.1'
}
...
```

此外，还可定义绑定表达式，考查下列示例：

```
<?xml version="1.0" encoding="utf-8"?>
<layoutxmlns:android="http://schemas.android.com/apk/res/android">

<data>
  <variable
    name="note"
    type="com.journaler.model.Note" />
</data>

<LinearLayout
  android:layout_width="match_parent"
    android:layout_height="match_parent"
    android:orientation="vertical">

    <TextView
        android:layout_width="wrap_content"
```

```
      android:layout_height="wrap_content"
      android:text="@{note.title}" />

  </LinearLayout>
</layout>
```

数据绑定如下所示。

```
package com.journaler.activity

import android.databinding.DataBindingUtil
import android.location.Location
import android.os.Bundle
import com.journaler.R
import com.journaler.databinding.ActivityBindingBinding
import com.journaler.model.Note

abstract class BindingActivity : BaseActivity() {

  override fun onCreate(savedInstanceState: Bundle?) {
    super.onCreate(savedInstanceState)
    /**
     * ActivityBindingBinding is auto generated class
     * which name is derived from activity_binding.xml filename.
     */
    val binding : ActivityBindingBinding =
    DataBindingUtil.setContentView(
        this, R.layout.activity_binding
    )
    val location = Location("dummy")
    val note = Note("my note", "bla", location)
    binding.note = note
  }

}
```

不难发现，数据与布局视图间的绑定十分简单，建议读者对此予以尝试。

12.12 使用列表

前述内容讲解了如何与数据协同工作。在主视图数据容器中，我们曾使用了

ListView——这是较为常用的加载数据的容器。在大多数时候，可采用 ListView 加载源自适配器的数据。注意，不要将大量的视图置于诸如 LinearLayout 这一类可滚动的容器中，并尽可能地使用 ListView。当不再使用视图时，ListView 可对其进行回收，并在需要时恢复视图。

使用列表可对应用程序性能产生影响，针对数据显示，它是一种经过良好优化的容器。在大多数应用程序中，显示列表是一个必备功能。作为操作结果，产生数据集的应用程序一般都会使用列表。

12.13 使用网格

12.12 节讨论了列表的重要性。如果将数据作为网格显示，情况又当如何？对此，Android Framework 提供了 GridView，其工作方式与 ListView 十分类似。在布局中用户可定义自己的 GridView，并将适配器实例分配于 GridView 的适配器属性中。GridView 将回收全部视图，并在必要时执行实例化操作。列表和网格间的主要差别在于，需要针对 GridView 定义列数。下列代码展示了 GridView 的应用示例：

```
<?xml version="1.0" encoding="utf-8"?>
<GridView xmlns:android="http://schemas.android.com/apk/res/android"
   android:id="@+id/my_grid"
   android:layout_width="match_parent"
   android:layout_height="match_parent"
   android:columnWidth="100dp"
   android:numColumns="3"
   android:verticalSpacing="20dp"
   android:horizontalSpacing="20dp"

   android:stretchMode="columnWidth"
   android:gravity="center"
/>
```

其中，一些较为重要的属性罗列如下：

- columnWidth：该属性定义了各列的宽度。
- numColumns：该属性指定了列数。
- verticalSpacing：该属性指定了行间的垂直间距。
- horizontalSpacing：该属性指定了网格条目间的水平间距。

尝试更新当前应用程序的主 ListView，以作为 GridView 表示数据。其间，可对其进行适当调整，以满足终端用户的要求。

12.14　实现拖曳操作

本节将展示如何实现拖曳特性。在大多数包含列表数据的应用程序中，该特定不可或缺。对于拖曳操作，使用列表并非强制行为。用户可以拖曳任何内容（视图），并将其释放到定义了适当监听器的任何地方。接下来将展示相关的实现示例。

首先定义一个视图，在该视图上，将设置一个长按监听器，并触发拖曳操作，如下所示。

```
view.setOnLongClickListener {
        val data = ClipData.newPlainText("", "")
        val shadowBuilder = View.DragShadowBuilder(view)
        view.startDrag(data, shadowBuilder, view, 0)
        true
}
```

此处使用了 ClipData 类传递数据以拖曳一个目标。下面定义 dragListener，并将其分配至一个我们期望它可以拖曳的视图上，如下所示。

```
private val dragListener = View.OnDragListener {
    view, event ->
    val tag = "Drag and drop"
    event?.let {
        when (event.action) {
            DragEvent.ACTION_DRAG_STARTED -> {
                Log.d(tag, "ACTION_DRAG_STARTED")
            }
            DragEvent.ACTION_DRAG_ENDED -> {
                Log.d(tag, "ACTION_DRAG_ENDED")
            }
            DragEvent.ACTION_DRAG_ENTERED -> {
                Log.d(tag, "ACTION_DRAG_ENDED")
            }
            DragEvent.ACTION_DRAG_EXITED -> {
                Log.d(tag, "ACTION_DRAG_ENDED")
            }
            else -> {
                Log.d(tag, "ACTION_DRAG_ ELSE ...")
            }
```

```
        }
    }
    true
  }

target?.setOnDragListener(dragListener)
```

当开始拖曳一个视图，并将其释放于 target 视图（分配了监听器）上时，拖曳监听器将触发上述代码。

12.15 本章小结

本章涉及了大量的主题。其间，我们学习了后端通信、如何利用 Retrofit 构建与后端远程实例间的通信，以及如何处理所得到的数据。本章的目标之一是与内容供应商和内容加载器协同工作。读者应了解这方面内容的重要性和优点。最后，本章还介绍了数据绑定机制，以及数据视图容器的重要性，例如 ListView 和 GridView，并展示了如何执行拖曳操作。第 13 章将对代码进行测试，以实现代码的优化。

第 13 章　性 能 调 优

第 12 章讨论了后端和 API 方面的内容，但我们的旅程仍未结束。本章将探讨性能优化方面的问题，并通过相关示例对此予以展示。具体来说，我们将考查相关代码，并遵循某些相关建议。

本章主要涉及以下主题：

- 布局优化。
- 电池寿命的优化。
- 获得最大的响应能力。

13.1　优 化 布 局

当获得最大的 UI 性能时，可遵循以下几点建议：

- 优化布局层次结构：应避免采用嵌套式布局，这会对性能产生严重的影响，如嵌套的多个 LinearLayout 视图。相反，可采用 RelativeLayout，这将显著地提升性能。在计算和绘制方面，嵌套布局往往会占用更多的处理资源。
- 尽可能地复用布局：对此，Android 提供了<include />。

考查下列示例：

```
to_be_included.xml:
<RelativeLayout xmlns:android=
  "http://schemas.android.com/apk/res/android"
  xmlns:tools="http://schemas.android.com/tools"
  android:layout_width="match_parent"
  android:layout_height="wrap_content"
  android:background="@color/main_bg"
  tools:showIn="@layout/includes" >

  <TextView
   android:id="@+id/title"
   android:layout_width="wrap_content"
   android:layout_height="wrap_content"
  />
```

```
</RelativeLayout>

includes.xml
  <LinearLayout xmlns:android=
  "http://schemas.android.com/apk/res/android"
   android:orientation="vertical"
   android:layout_width="match_parent"
   android:layout_height="match_parent"
   android:background="@color/main_bg"
  >
  ...
  <include layout="@layout/to_be_included"/>
  ...
</LinearLayout>
```

- 除此之外，还可使用<merge>。在将一个布局包含在另一个布局中时，<merge>消除了视图层次结构中的冗余视图分组。考查下列示例：

```
to_merge.xml
<merge xmlns:android="http://schemas.android.com/apk/res/android">

 <ImageView
   android:id="@+id/first"
   android:layout_width="fill_parent"
   android:layout_height="wrap_content"
   android:src="@drawable/first"/>

 <ImageView
   android:id="@+id/second"
   android:layout_width="fill_parent"
   android:layout_height="wrap_content"
   android:src="@drawable/second"/>

</merge>
```

当利用 include 在另一个布局中包含 to_merge.xml 时，Android 将忽略<merge>，并将视图直接添加至<include />所设置的容器中，具体如下：

- 仅当需要时将布局包含至屏幕中——如果当前不需要使用视图，可将其可见性设置为 Gone，而非 Invisible。相应地，Invisible 将创建一个视图实例。当使用 Gone 时，Android 仅在可见性修改为 Visible 时实例化视图。
- 使用 ListView 或 GridView 加载数据分组，第 12 章曾对此有所解释。

13.2　优化电池寿命

电池损耗涉及多方面的原因。原因之一是在应用程序中执行了过多的操作，这将对电池电量带来显著的影响。本节将讨论节省电量的各种建议。

为了使电池处于最佳状态，应注意以下几点内容：

- 尽可能地减少网络通信。频繁的网络调用将会影响电池的电量，因而应尽量避免。
- 确定手机是否正在充电。此时，应用程序可执行高强度和高性能的操作。
- 监视连接状态。只有在连接状态正常时才执行与连接相关的操作。
- 合理利用广播信息。频繁且不必要地发送广播消息会对性能产生影响。
- 应注意 GPS 的使用强度。频繁的地理位置请求可对电池电量带来显著的影响。

13.3　保持应用程序响应性

相信读者都会遇到“应用程序无响应”这一类消息。针对于此，本节罗列出以下几点建议：

- 确保输入未被阻塞（任何密集型操作，特别是网络流量）。
- 不要在主应用程序线程上执行较为耗时的任务。
- 不要在 onReceive()方法中为广播接收器执行长时间运行的操作。
- 尽量使用 AsyncTask 类。另外，还可考虑使用 ThreadPoolExecutor。
- 尽可能地使用内容加载器。
- 避免同时执行过多的线程。
- 在独立线程中执行文件系统的写入操作。

如果仍出现 ANR 等问题，或者应用程序依然运行缓慢，那么可以使用 Systrace 和 Traceview 等工具来跟踪问题的根源。

13.4　本 章 小 结

本章篇幅较为简短，主要讨论了如何维护和实现应用程序的良好性能和响应性，相关建议在应用程序优化过程中十分重要。如果应用程序尚未对此予以涉及，建议以此执行相应的优化操作。第 14 章将对应用程序进行测试，其中包括单元测试和测量测试。

第 14 章　测　　试

前述章节利用代码库开发了一个应用程序，且认为该程序不存在 Bug。然而，这一结论可能并不正确。即使我们确信应用程序中不含有 Bug，但依然会存在某些问题。对此，需要编写测试以检测代码。本章将通过相关示例展示如何设置、编写和运行测试程序。

本章主要涉及以下主题：

- 如何编写第一个测试程序。
- 使用测试套件。
- 如何测试 UI。
- 运行测试。
- 单元和设备测试。

14.1　添加依赖关系

当运行测试时，需要设置相关的依赖关系。下面将通过扩展 build.gradle 更新应用程序配置，进而支持测试行为并提供所需的类。对此，打开 build.gradle 并按照下列方式进行扩展：

```
apply plugin: "com.android.application"
apply plugin: "kotlin-android"
apply plugin: "kotlin-android-extensions"

repositories {
  maven { url "https://maven.google.com" }
}

android {
  ...
  sourceSets {
    main.java.srcDirs += [
            'src/main/kotlin',
            'src/common/kotlin',
            'src/debug/kotlin',
            'src/release/kotlin',
```

```
            'src/staging/kotlin',
            'src/preproduction/kotlin',
            'src/debug/java',
            'src/release/java',
            'src/staging/java',
            'src/preproduction/java',
            'src/testDebug/java',
            'src/testDebug/kotlin',
            'src/androidTestDebug/java',
            'src/androidTestDebug/kotlin'
    ]
  }
  ...
  testOptions {
    unitTests.returnDefaultValues = true
  }
}
...
dependencies {
  ...
  compile "junit:junit:4.12"
  testCompile "junit:junit:4.12"

  testCompile "org.jetbrains.kotlin:kotlin-reflect:1.1.51"
  testCompile "org.jetbrains.kotlin:kotlin-stdlib:1.1.51"

  compile "org.jetbrains.kotlin:kotlin-test:1.1.51"
  testCompile "org.jetbrains.kotlin:kotlin-test:1.1.51"

  compile "org.jetbrains.kotlin:kotlin-test-junit:1.1.51"
  testCompile "org.jetbrains.kotlin:kotlin-test-junit:1.1.51"

  compile 'com.android.support:support-annotations:26.0.1'
  androidTestCompile 'com.android.support:support
  -annotations:26.0.1'

  compile 'com.android.support.test:runner:0.5'
  androidTestCompile 'com.android.support.test:runner:0.5'

  compile 'com.android.support.test:rules:0.5'
  androidTestCompile 'com.android.support.test:rules:0.5'
}
```

```
It is important to highlight use of:
testOptions {
    unitTests.returnDefaultValues = true
}
```

据此，可对内容供应商进行测试，并在测试中使用全部相关类；否则，将得到下列错误消息：

```
Error: "Method ... not mocked"!
```

14.2　更新文件夹结构

文件夹结构以及其中的代码必须遵循有关构建变体的约定。针对当前测试，将使用下列结构部分：

- ❑ 单元测试如图 14.1 所示。
- ❑ 设备测试如图 14.2 所示。

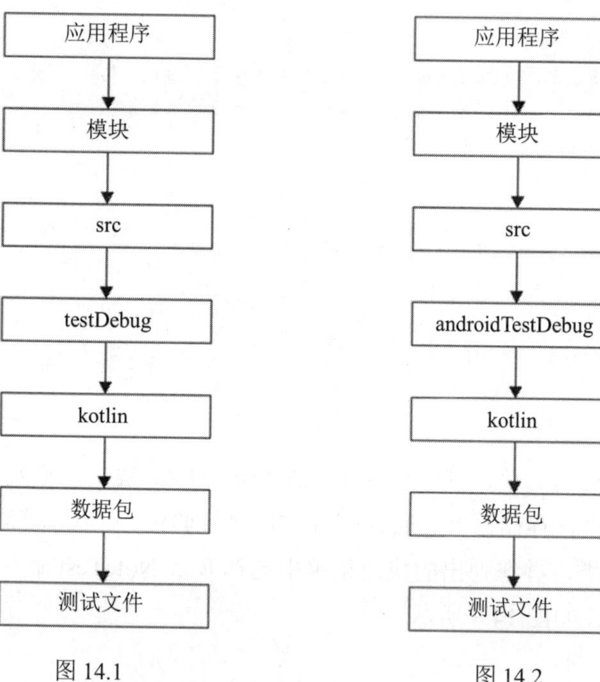

图 14.1　　　　图 14.2

下面开始编写测试代码。

14.3　编写第一个测试

在单元测试的 root 数据包中，定义一个名为 NoteTest 的新类，如下所示。

```
package com.journaler

import android.location.Location
import com.journaler.database.Content
import com.journaler.model.Note
import org.junit.Test

class NoteTest {

  @Test
  fun noteTest() {
    val note = Note(
            "stub ${System.currentTimeMillis()}",
            "stub ${System.currentTimeMillis()}",
            Location("Stub")
    )

    val id = Content.NOTE.insert(note)
    note.id = id

    assert(note.id > 0)
  }
}
```

上述测试过程较为简单，其中创建了一个新的 Note 实例，触发了内容供应商中的 CRUD 操作以对其进行存储，并对所接收到的 ID 进行验证。当运行该测试时，右击 Project 面板中的 NoteTest 类，并在弹出的快捷菜单中选择 Run 'NoteTest'命令，如图 14.3 所示。

执行单元测试，如图 14.4 所示。

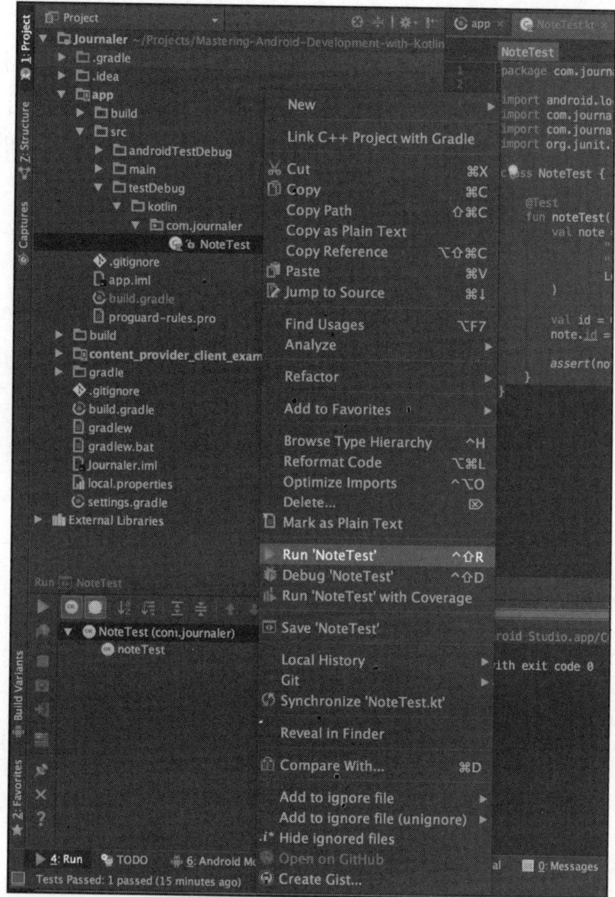

图 14.3

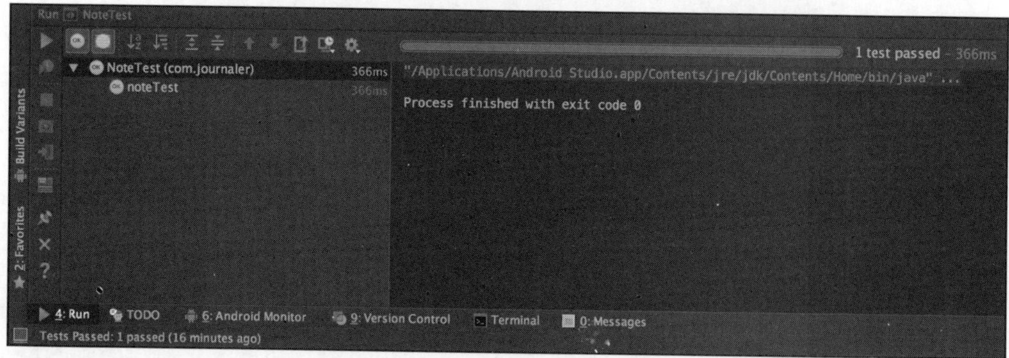

图 14.4

可以看到，此处成功地向数据库中插入了 Note。在构建了第一个单元测试后，接下来将创建第一个设备测试。设备测试与单元测试间的差别在于，设备测试运行于设备或模拟器上。当需要测试依赖于 Android Context 的代码时，可以使用它们。下面将测试 main 服务，对此，在设备测试 root 数据包中定义一个名为 MainServiceTest 的新类，如下所示。

```
package com.journaler

import android.content.ComponentName
import android.content.Context
import android.content.Intent
import android.content.ServiceConnection
import android.os.IBinder
import android.support.test.InstrumentationRegistry
import android.util.Log
import com.journaler.service.MainService
import org.junit.After
import org.junit.Before
import org.junit.Test
import kotlin.test.assertNotNull

class MainServiceTest {

  private var ctx: Context? = null
  private val tag = "Main service test"

  private val serviceConnection = object : ServiceConnection {
    override fun onServiceConnected(p0: ComponentName?, binder:
    IBinder?) {
      Log.v(tag, "Service connected")
    }

    override fun onServiceDisconnected(p0: ComponentName?) {
      Log.v(tag, "Service disconnected")
    }
  }

  @Before
  fun beforeMainServiceTest() {
    Log.v(tag, "Starting")
    ctx = InstrumentationRegistry.getInstrumentation().context
  }

  @Test
  fun testMainService() {
    Log.v(tag, "Running")
```

```
    assertNotNull(ctx)
    val serviceIntent = Intent(ctx, MainService::class.java)
    ctx?.startService(serviceIntent)
    val result = ctx?.bindService(
      serviceIntent,
      serviceConnection,
      android.content.Context.BIND_AUTO_CREATE
    )
    assert(result != null && result)
  }

  @After
  fun afterMainServiceTest() {
    Log.v(tag, "Finishing")
    ctx?.unbindService(serviceConnection)
    val serviceIntent = Intent(ctx, MainService::class.java)
    ctx?.stopService(serviceIntent)
  }

}
```

运行后将生成如图 14.5 所示的新配置。

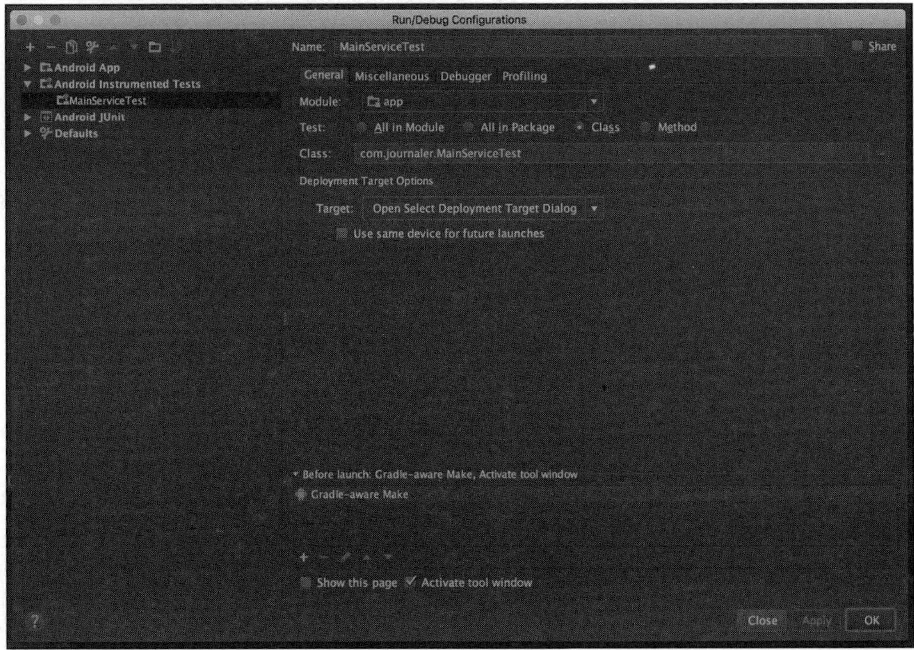

图 14.5

运行新创建的配置时，将会被询问选择运行测试的 Android 设备或模拟器实例，如图 14.6 所示。

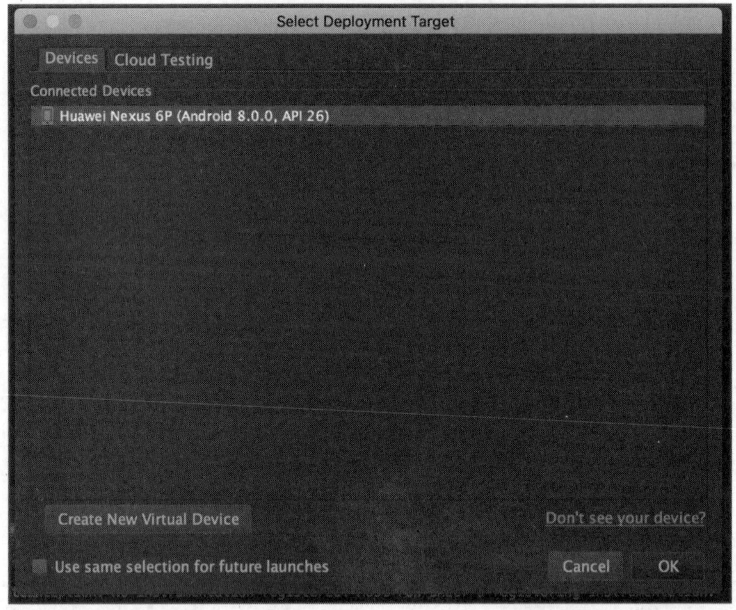

图 14.6

至此，我们成功地创建并运行了设备测试。当读者尝试练习时，可尽可能地定义多个测试，覆盖应用程序所包含的全部代码，并留意测试过程是单元测试还是设备测试。

14.4　使用单元测试套件

测试套件表示为一个测试集合，本节将展示如何创建测试集合。下面创建一个测试并表示为集合容器，同时将其命名为 MainSuite，如下所示。

```
package com.journaler

import org.junit.runner.RunWith
import org.junit.runners.Suite

@RunWith(Suite::class)
@Suite.SuiteClasses(
    DummyTest::class,
```

```
    MainServiceTest::class
)
class MainSuite
```

重复设备测试中的各项步骤，进而运行测试套件。

测试 UI 可以防止出现意外情况、应用程序崩溃或性能低下。

因此，这里也强烈建议编写 UI 测试，以确保 UI 按照期望方式执行。对此，我们将引入 Espresso Framework。

首先对其添加依赖关系，如下所示。

```
...
compile 'com.android.support.test.espresso:espresso-core:2.2.2'
androidTestCompile 'com.android.support.test.espresso:espressocore:
2.2.2'
...
```

在编写和运行 Espresso 测试之前，需要禁用测试设备上的动画功能，其原因在于，这将对测试过程、期望事件以及相关行为产生影响。相应地，访问设备的 Settings | Developer options |，并关闭下列功能项：

- Window animation scale。
- Transition animation scale。
- Animator duration scale。

接下来开始编写 Espresso 测试，并考查以下示例：

```
@RunWith(AndroidJUnit4::class)
class MainScreenTest {
    @Rule
    val mainActivityRule =
    ActivityTestRule(MainActivity::class.java)

    @Test
    fun testMainActivity(){
      onView((withId(R.id.toolbar))).perform(click())
      onView(withText("My dialog")).check(matches(isDisplayed()))
   }
}
```

这里将对 MainActivity 类进行测试。在测试触发了工具栏按钮单击操作后，将查看是否弹出对话框。此处，我们通过检查标签“My dialog”的有效性来实现这一点。Espresso Framework 的整体内容超出了本书的讨论范围，当前仅展示了少许提示性内容。

14.5　运行测试

前述内容通过 Android Studio 执行了相关测试。当测试编写完毕后，通常希望一次性地运行它们。针对所有的构建变化版本，可以运行全部的单元测试，但只能针对特定的风格或构建类型。该过程同样适用于设备测试，稍后将展示相应的示例，并针对 Journaler 应用程序使用现有的构建变化版本实现这一任务。

14.5.1　运行单元测试

打开终端并访问当前项目的 root 数据包。当运行全部单元测试时，可执行下列命令行：

```
$ ./gtradlew test
```

这将运行之前所编写的全部单元测试。由于 NoteTest 使用了内容供应商提供的内容，因而测试过程将失败。针对于此，需要通过相应的 Runner 类予以执行。默认状态下，Android Studio 负责完成这项工作。然而，由于这是一个单元测试，并且将从终端对其加以执行，因而测试将会失败。

由于使用了 Android Framework 组件，实际上该测试可视为一种测试设备。常见的做法是，如果类依赖于 Android Framework 组件，则需要作为测试设备予以执行。因此，可将 NoteTest 移至一个设备测试目录中。相应地，当前不存在任何单元测试。创建至少一个不依赖于 Android Framework 组件的组件，基于此，可将已有的 DummyTest 移至单元测试文件夹中。从 IDE 中拖放它，然后采用相同的命令重新运行测试。

当针对构建变化版本运行全部测试时，可执行下列命令行：

```
$ ./gradlew testCompleteDebug
```

这里，我们针对 Complete 风格和 Debug 构建类型执行了测试。

14.5.2　运行设备测试

当运行设备测试时，可采用以下命令行：

```
$ ./gradlew connectedAndroidTest
```

其先决条件是设备处于连接状态，或模拟器处于运行状态。如果存在多台设备或模拟器，它们都将运行测试。

当针对构建变化版本运行设备测试时，可使用下列命令行：

```
$ ./gradlew connectedCompleteDebugAndroidTest
```

上述命令行表示将利用 Debug 构建类型并针对 connected 风格触发全部设备测试。

14.6　本章小结

本章讨论了如何针对应用程序编写和运行测试，这也是迈向最终产品的重要一步，进而构建了编写良好且不存在任何 Bug 的产品。稍后，我们将讨论如何发布产品。

第 15 章　迁移至 Kotlin

如果遗留项目或现有 Java 模块需要迁移至 Kotlin 中，实际上该过程并不复杂。回忆一下，Kotlin 具有互操作性，因此，一些模块不需要完全迁移；相反，它们可以包含在 Kotlin 项目中。当然，具体决定权掌握在用户手中。

本章主要涉及以下主题：

- 迁移的准备工作。
- 转换类。
- 重构和清除。

15.1　迁移的准备工作

如前所述，我们需要做出决定：是否采用 Kotlin 完全重写模块；或者是继续采用 Kotlin 编写代码，但将遗留代码保持为纯 Java。

当前项目尚未包含任何可迁移内容。因此，我们将创建一些代码。如果尚未设置包含数据包结构的 Java 源目录，则需要对其加以构建。下面添加下列数据包：

- activity。
- model。

此类数据包等同于 Kotlin 源代码中已有的数据包。在 activity 数据包中，添加下列类：

- MigrationActivity.java 代码如下所示。

```
package com.journaler.activity;

import android.os.Bundle;
import android.support.annotation.Nullable;
import android.support.v7.app.AppCompatActivity;

import com.journaler.R;

public class MigrationActivity extends AppCompatActivity {

  @Override
  protected void onCreate(@Nullable Bundle savedInstanceState) {
```

```
    super.onCreate(savedInstanceState);
    setContentView(R.layout.activity_main);
  }

  @Override
    protected void onResume() {
    super.onResume();
  }
}
```

- MigrationActivity2.java：确保与 MigrationActivity.java 包含相同的实现。这里仅需要一些代码库来呈现和迁移。随后，将两项活动注册至 Android manifest 文件中，如下所示。

```
<manifest xmlns:android=
  "http://schemas.android.com/apk/res/android"
  package="com.journaler">
  ...
  <application
    ...
  >
  ...
    <activity
      android:name=".activity.MainActivity"
      android:configChanges="orientation"
      android:screenOrientation="portrait">
      <intent-filter>
        <action android:name="android.intent.action.MAIN" />
        <category android:name=
        "android.intent.category.LAUNCHER" />
      </intent-filter>
    </activity>

    <activity
      android:name=".activity.NoteActivity"
      android:configChanges="orientation"
      android:screenOrientation="portrait" />

    <activity
      android:name=".activity.TodoActivity"
      android:configChanges="orientation"
      android:screenOrientation="portrait" />
```

```
    <activity
      android:name=".activity.MigrationActivity"
      android:configChanges="orientation"
      android:screenOrientation="portrait" />

    <activity
      android:name=".activity.MigrationActivity2"
      android:configChanges="orientation"
      android:screenOrientation="portrait" />
  </application>
</manifest>
```

可以看到，Java 代码与 Kotlin 代码共处，且不会产生任何问题，当前 Android 项目均可对二者加以使用。接下来向 model 数据包中添加下列类：

- Dummy.java 代码如下所示。

```
package com.journaler.model;

public class Dummy {

  private String title;
  private String content;

  public Dummy(String title) {
    this.title = title;
  }

  public Dummy(String title, String content) {
    this.title = title;
    this.content = content;
  }

  public String getTitle() {
    return title;
  }

  public void setTitle(String title) {
    this.title = title;
  }

  public String getContent() {
    return content;
  }
```

```
  public void setContent(String content) {
    this.content = content;
  }

}
```

- Dummy2.java 代码如下所示。

```
package com.journaler.model;

import android.os.Parcel;
import android.os.Parcelable;

public class Dummy2 implements Parcelable {

    private int count;
    private float result;

    public Dummy2(int count) {
      this.count = count;
      this.result = count * 100;
  }

  public Dummy2(Parcel in) {
    count = in.readInt();
    result = in.readFloat();
  }

  public static final Creator<Dummy2>
  CREATOR = new Creator<Dummy2>() {
    @Override
    public Dummy2 createFromParcel(Parcel in) {
      return new Dummy2(in);
    }
    @Override
    public Dummy2[] newArray(int size) {
      return new Dummy2[size];
    }
  };

  @Override
  public void writeToParcel(Parcel parcel, int i) {
```

```
    parcel.writeInt(count);
    parcel.writeFloat(result);
  }

  @Override
  public int describeContents() {
    return 0;
  }

  public int getCount() {
    return count;
  }

  public float getResult() {
    return result;
  }
}
```

下面再次检查项目的 Kotlin 部分是否能够看到这些类。在 Kotlin 源目录的根目录中创建一个新的.kt 文件，并将其命名为 kotlin_calls_java.kt，如下所示。

```
package com.journaler

import android.content.Context
import android.content.Intent
import com.journaler.activity.MigrationActivity
import com.journaler.model.Dummy2

fun kotlinCallsJava(ctx: Context) {

  /**
  * We access Java class and instantiate it.
  */
  val dummy = Dummy2(10)

  /**
  * We use Android related Java code with no problems as well.
  */
    val intent = Intent(ctx, MigrationActivity::class.java)
    intent.putExtra("dummy", dummy)
    ctx.startActivity(intent)
}
```

不难发现，当采用 Java 代码时，Kotlin 不会产生任何问题。所以仍可继续执行迁移操作，稍后将对此加以展示。

15.2 危险信号

将庞大而复杂的 Java 类转换为 Kotlin 仍然是一种选择方案。无论如何，我们需要提供适当的单元测试或设备测试，以便在转换之后重新测试这些类的功能。如果测试失败，需要再次检查失败的原因。

相应地，需要迁移的类可按照以下两种方案进行：

- 自动化转换。
- 手动重写。

对于庞大而较为复杂的类，上述两种方案均包含了某些缺陷。有些时候，完全自动化可能会生成不完美的代码，随后还需要对其再次进行检查和格式化。另外，第二种方案则较为耗时。

最终结论是，总是可以使用原始的 Java 代码。从切换至 Kotlin 作为主要语言开始，即可在 Kotlin 中编写所有全新的内容。

15.3 更新依赖关系

如果打算将 Android 项目的 100%纯 Java 代码切换至 Kotlin 中，则必须从头开始完成这一过程。这意味着，第一次迁移将更新依赖关系。对此，必须更改 build.gradle 配置，以便 Kotlin 被识别且源代码路径有效，这已在第 1 章中有所解释。因此，如果项目尚未包含与 Kotlin 相关的配置，则必须对此进行更新。

下面简要描述一下 Gradle 的配置过程。

- build.gradle 根项目须体现 build.gradle 主文件，如下所示。

```
buildscript {
  repositories {
    jcenter()
    mavenCentral()
  }
  dependencies {
    classpath 'com.android.tools.build:gradle:2.3.3'
    classpath 'org.jetbrains.kotlin:kotlin-gradle-plugin: 1.1.51'
```

```
  }
}

repositories {
  jcenter()
  mavenCentral()
}
```

build.gradle 主应用程序负责处理应用程序的所有依赖关系，如下所示。

```
apply plugin: "com.android.application"
apply plugin: "kotlin-android"
apply plugin: "kotlin-android-extensions"

repositories {
  maven { url "https://maven.google.com" }
}

android {
  ...
  sourceSets {
    main.java.srcDirs += [
        'src/main/kotlin',
        'src/common/kotlin',
        'src/debug/kotlin',
        'src/release/kotlin',
        'src/staging/kotlin',
        'src/preproduction/kotlin',
        'src/debug/java',
        'src/release/java',
        'src/staging/java',
        'src/preproduction/java',
        'src/testDebug/java',
        'src/testDebug/kotlin',
        'src/androidTestDebug/java',
        'src/androidTestDebug/kotlin'
    ]
    ...
  }
  ...
}

repositories {
```

```
  jcenter()
  mavenCentral()
}

dependencies {
  compile "org.jetbrains.kotlin:kotlin-reflect:1.1.51"
  compile "org.jetbrains.kotlin:kotlin-stdlib:1.1.51"
  ...
  compile "com.github.salomonbrys.kotson:kotson:2.3.0"
  ...

  compile "junit:junit:4.12"
  testCompile "junit:junit:4.12"

  testCompile "org.jetbrains.kotlin:kotlin-reflect:1.1.51"
  testCompile "org.jetbrains.kotlin:kotlin-stdlib:1.1.51"

  compile "org.jetbrains.kotlin:kotlin-test:1.1.51"
  testCompile "org.jetbrains.kotlin:kotlin-test:1.1.51"

  compile "org.jetbrains.kotlin:kotlin-test-junit:1.1.51"
  testCompile "org.jetbrains.kotlin:kotlin-test-junit:1.1.51"
  ...
}
```

上述内容展示了须满足的所有与 Kotlin 相关的依赖关系。其中之一是 Kotson，并提供了针对 Gson 库的 Kotlin 绑定。

15.4　转　换　类

最后是对类进行迁移。对此存在两种自动选择方案，本节将对二者加以使用。首先，找到 MigrationActivity.java 文件并将其打开；然后，选择 Code | Convert Java File To Kotlin File，其间，转换过程可能会花费几秒时间；接下来，将该文件从 Java 数据包中拖曳至 Kotlin 源数据包中。随后，观察下列源代码：

```
package com.journaler.activity

import android.os.Bundle
import android.support.v7.app.AppCompatActivity
```

```
import com.journaler.R

class MigrationActivity : AppCompatActivity() {

  override fun onCreate(savedInstanceState: Bundle?) {
    super.onCreate(savedInstanceState)
    setContentView(R.layout.activity_main)
  }

  override fun onResume() {
    super.onResume()
  }

}
```

如前所述，全自动转换有时不会生成完美的代码，稍后将会执行重构和清理工作。在第二种方案中，完成相同任务则需要将 Java 代码复制、粘贴至 Kotlin 文件中，并从 MigrationActivity2 中复制全部源代码。随后，创建包含相同名称的一个新的 Kotlin 类，并粘贴代码。当被询问时，可选择执行自动转换。在代码出现后，可移除该类的 Java 版本。通过观察可知，源代码与迁移的 MigrationActivity 类的源代码是相同的。

接下来针对 Dummy 类和 Dummy2 类实施上述两种方案。

- Dummy 类代码如下所示。

```
package com.journaler.model

class Dummy {

  var title: String? = null
  var content: String? = null

  constructor(title: String) {
    this.title = title
  }

  constructor(title: String, content: String) {
    this.title = title
    this.content = content
  }

}
```

- Dummy2 类代码如下所示。

```
package com.journaler.model

import android.os.Parcel
import android.os.Parcelable

class Dummy2 : Parcelable {

  var count: Int = 0
  private set
  var result: Float = 0.toFloat()
  private set
  constructor(count: Int) {
    this.count = count
    this.result = (count * 100).toFloat()
  }

  constructor('in': Parcel) {
    count = 'in'.readInt()
    result = 'in'.readFloat()
  }

  override fun writeToParcel(parcel: Parcel, i: Int) {
    parcel.writeInt(count)
    parcel.writeFloat(result)
  }

  override fun describeContents(): Int {
    return 0
  }

  companion object {

    val CREATOR: Parcelable.Creator<Dummy2>
    = object : Parcelable.Creator<Dummy2> {
       override fun createFromParcel('in': Parcel): Dummy2 {
         return Dummy2('in')
       }

       override fun newArray(size: Int): Array<Dummy2> {
         return arrayOfNulls(size)
       }
    }
  }

}
```

其中，Dummy2 类包含了某些与转换相关的问题，且必须亲自对其进行修复。对应问题出现在以下代码行中：

```
override fun newArray(size: Int): Array<Dummy2> { ...
```

对此，可将 Array<Dummy2> int Array<Dummy2?>转换为以下类型：

```
override fun newArsray(size: Int): Array<Dummy2?> { ...
```

这也是在进行迁移时可能面临的挑战！值得注意的是，在 Dummy 类和 Dummy2 类中，我们都通过切换到 Kotlin 中而显著地减少了代码库。由于不再有 Java 实现，因此可以进行代码重构和清理。

15.5 重构和清理

为了在转换后获得较优的代码，必须执行相应的重构和清理操作。对此，可调整代码库以符合 Kotlin 标准和习惯用法，并通读全部内容。只有当这一切完成后，方可认为迁移已经完成。

打开相关类并仔细阅读代码，其中仍存在诸多改进空间。在适当修正后，MigrationActivity 代码如下所示。

```
...
override fun onResume() = super.onResume()
...
```

不难发现，MigrationActivity（以及 MigrationActivity2）的工作量较少，这两个类均为小型类；而 Dummy 类和 Dummy2 类则涉及较大的工作量。

- Dummy 类的代码如下所示。

```
package com.journaler.model

class Dummy(
  var title: String,
  var content: String
) {

  constructor(title: String) : this(title, "") {
    this.title = title
  }

}
```

- Dummy2 类的代码如下所示。

```
package com.journaler.model

import android.os.Parcel
import android.os.Parcelable

class Dummy2(
 private var count: Int
) : Parcelable {

 companion object {
   val CREATOR: Parcelable.Creator<Dummy2>
   = object : Parcelable.Creator<Dummy2> {
     override fun createFromParcel(`in`: Parcel):
     Dummy2 = Dummy2(`in`)
     override fun newArray(size: Int): Array<Dummy2?> =
     arrayOfNulls(size)
   }
 }

 private var result: Float = (count * 100).toFloat()

 constructor(`in`: Parcel) : this(`in`.readInt())

 override fun writeToParcel(parcel: Parcel, i: Int) {
   parcel.writeInt(count)
 }

 override fun describeContents() = 0

}
```

与转换后的第一个 Kotlin 版本相比，重构后的这两个类版本得到了较大的改进。尝试将当前版本与所持有的原始 Java 代码进行比较，并得出自己的结论。

15.6　本章小结

本章讨论了如何实现 Kotlin 的迁移操作。关于迁移的方式和时机，本章介绍了相关技术和建议。可以看到，具体过程并不十分复杂。第 16 章将讨论应用程序的发布操作。

第 16 章　部署应用程序

现在，是时候向世界展示你的作品了。对此，本章将执行某些准备工作，最终将应用程序发布到 Google Play store 中。

本章主要涉及以下主题：

- 代码混淆技术。
- 签署应用程序。
- 部署至 Google Play 中。

16.1　部署的准备工作

在发布应用程序之前，需要执行某些准备工作。首先，需要移除不再使用的资源或类；随后还需要禁用日志机制。一种较好的方法是使用一些主流日志库。相应地，可围绕 Log 类创建一个封装器，针对每个日志输出设置一个条件，并检测不为 release 构建类型。

如果还未将发布配置设置为可调试的，可参照以下步骤进行：

```
...
buildTypes {
  ...
  release {
    debuggable false
  }
}
...
```

随后，可再次检测 manifest 并对其清空，移除任何不再需要的授权验证。具体来说，须移除以下内容：

```
<uses-permission android:name="android.permission.VIBRATE" />
```

之前曾添加了上述内容，但从未对其加以使用。最后一步则是检查应用程序的兼容性；查看最小和最大 SDK 版本是否符合设备的目标规划。

16.2　代码混淆技术

发布过程的下一项工作时启用代码混淆。对此，打开 build.gradle 配置，并按照下列方式进行更新：

```
...
buildTypes {
  ...
  release {
    debuggable false
    minifyEnabled true
    proguardFiles getDefaultProguardFile('proguard-android.txt'),
     'proguard-rules.pro'
  }
}
...
```

上述添加的配置将收缩资源并执行混淆操作。对此，这里使用了 ProGuard。ProGuard 是一个免费的 Java 类文件收缩器、优化器、混淆器和预校验器，并对未使用的类、字段、方法和属性进行检测，同时还将对字节码进行优化。

大多数时候，当移除全部未使用的代码时，默认的 ProGuard 配置已然足够；但有些时候，也可根据应用程序的实际需要定义 ProGuard 配置。针对于此，需要定义 ProGuard 配置来保存相关类。下面打开 ProGuard 配置文件，并添加下列内容：

```
-keep public class MyClass
```

以下内容是使用一些库时需要添加的 ProGuard 指令列表。

- Retorfit：

```
-dontwarn retrofit.**
-keep class retrofit.** { *; }
-keepattributes Signature
-keepattributes Exceptions
```

- Okhttp3：

```
-keepattributes Signature
-keepattributes *Annotation*
-keep class okhttp3.** { *; }
-keep interface okhttp3.** { *; }
-dontwarn okhttp3.**
-dontnote okhttp3.**

# Okio
-keep class sun.misc.Unsafe { *; }
-dontwarn java.nio.file.*
-dontwarn org.codehaus.mojo.animal_sniffer.IgnoreJRERequirement
```

❑ Gson：

```
-keep class sun.misc.Unsafe { *; }
-keep class com.google.gson.stream.** { *; }
```

随后，可以此更新 proguard-rules.pro 文件。

16.3　签署应用程序

在将发布版本上传至 Google Play store 中之前，最后一步是生成签署的 APK。对此，打开当前项目并选择 Build | Generate Signed APK，则出现如图 16.1 所示的页面。

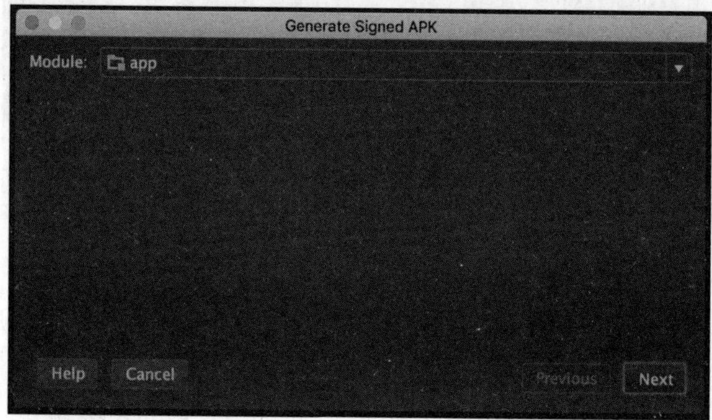

图 16.1

在图 16.1 中，在 Module 下拉列表框中选择主应用程序模块并单击 Next 按钮，则出现如图 16.2 所示的页面。

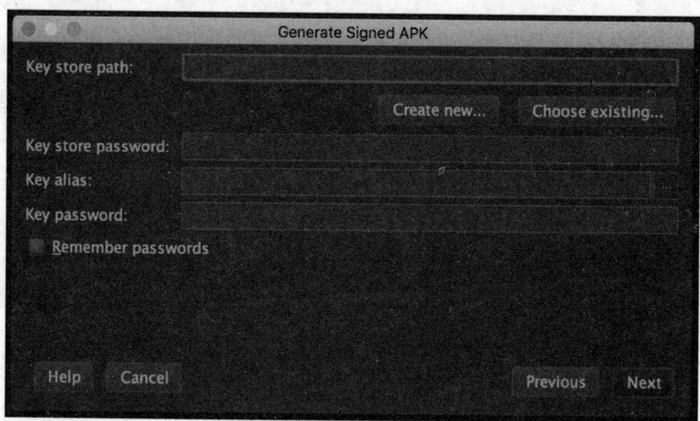

图 16.2

由于尚未存储任何新的密钥，此处将创建一个新的密钥。单击 Create new...按钮，则出现如图 16.3 所示的页面。

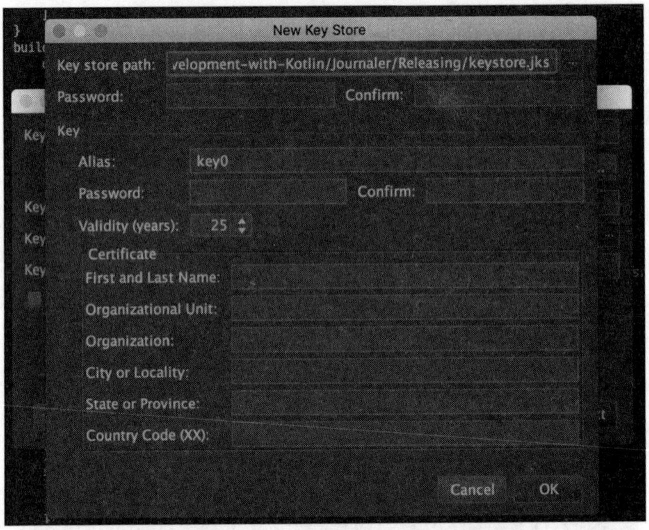

图 16.3

在图 16.3 中，填充相关数据并单击 OK 按钮，此时返回图 16.2 所示的页面，单击 Next 按钮并在后续询问过程中输入新的密码。随后在出现新的页面中，检查两种签名，并选择 complete 为构建风格，如图 16.4 所示，然后单击 Finish 按钮。

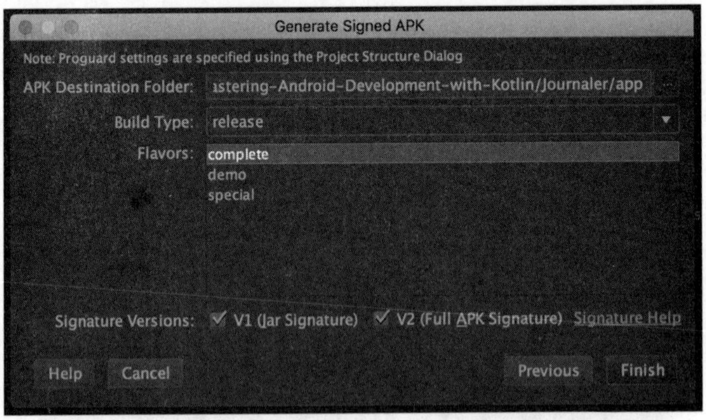

图 16.4

稍作等待直至构建过程完毕。此外，还将更新 build.gradle 文件，以便每次构建一个版本时均对其进行签名，如下所示。

```
...
android {
  signingConfigs {
    release {
      storeFile file("Releasing/keystore.jks")
      storePassword "1234567"
      keyAlias "key0"
      keyPassword "1234567"
    }
  }
  release {
    debuggable false
    minifyEnabled false
    signingConfig signingConfigs.release
    proguardFiles getDefaultProguardFile('proguard-android.txt'),
    'proguard-rules.pro'
  }
}
...
```

另外，还可在终端中运行构建处理过程，如下所示。

```
$ ./gradlew clean
$ ./gradlew assembleCompleteRelease
```

此处针对 Complete 应用程序风格构建了发布版本。

16.4　发布至 Google Play 中

部署过程中的最后一步是发布签署后的 APK。除了 APK 之外，还需要提供以下内容：

- 准备应用程序的截屏。在 Android Studio Logcat 中，单击 Screen Capture 图标（相机图标）。在 Preview 窗口中，单击 Save 按钮。随后将被询问是否保存图像，如图 16.5 所示。
- 包含以下规格的高分辨率图标：
    - 32 位的 PNG 图像（包含 Alpha 值）。
    - 512 像素×512 像素。
    - 1024KB 的最大文件尺寸。
- 功能图（应用程序的主 Banner）。
    - 24 位的 JPEG 图像或 PNG 图像（不包含 Alpha 值）。

➢ 1024 像素×500 像素。

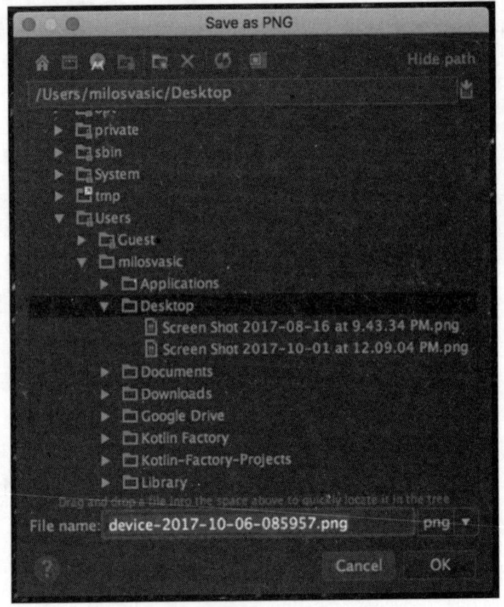

图 16.5

- 如果将应用程序发布为 TV 应用程序或 TV Banner：
  ➢ 24 位的 JPEG 图像或 PNG 图像（不包含 Alpha 值）。
  ➢ 1280 像素×720 像素。
- 宣传视频——YouTube 视频（不是播放列表）。
- 应用程序的纹理描述。

登录控制台（https://play.google.com/apps/publish）。如果尚未注册，应先行完成注册过程。图 16.6 显示了主控制台页面。

当前尚未发布任何应用程序。单击 PUBLISH AN ANDROID APP ON GOOGLE PLAY，随后将弹出 Create application 对话框，填写相关数据并单击 CREATE 按钮，如图 16.7 所示。

填充表单数据，如图 16.8 所示。

更新图形资源数据，如图 16.9 所示。

查看如图 16.10 所示的内容。

应用程序分类如图 16.11 所示。

联系方式和隐私策略如图 16.12 所示。

当填写完所有的必要数据后，滚动回页面上方并单击 SAVE DRAFT 按钮；在页面左侧选择 App releases，随后将显示如图 16.13 所示的页面。

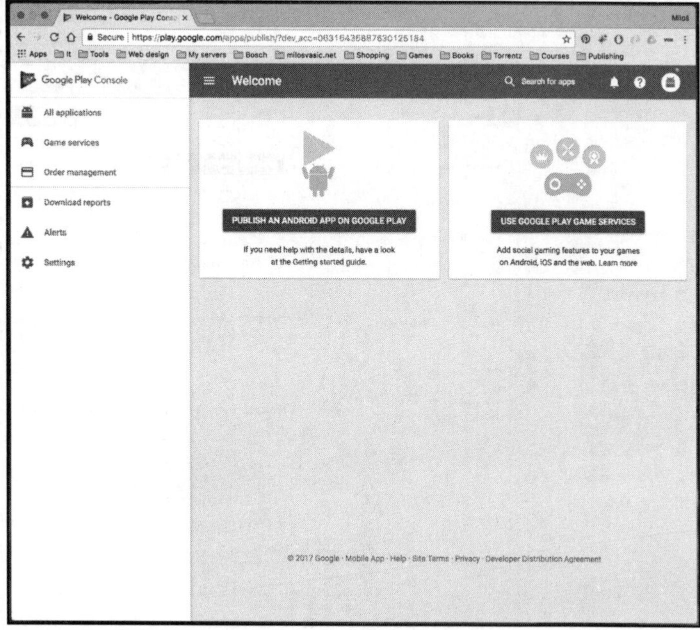
Welcome - Google Play Console
Google Play Console
Welcome
All applications
Game services
Order management
Download reports
Alerts
Settings
PUBLISH AN ANDROID APP ON GOOGLE PLAY
If you need help with the details, have a look at the Getting started guide.
USE GOOGLE PLAY GAME SERVICES
Add social gaming features to your games on Android, iOS and the web. Learn more
© 2017 Google · Mobile App · Help · Site Terms · Privacy · Developer Distribution Agreement

图 16.6

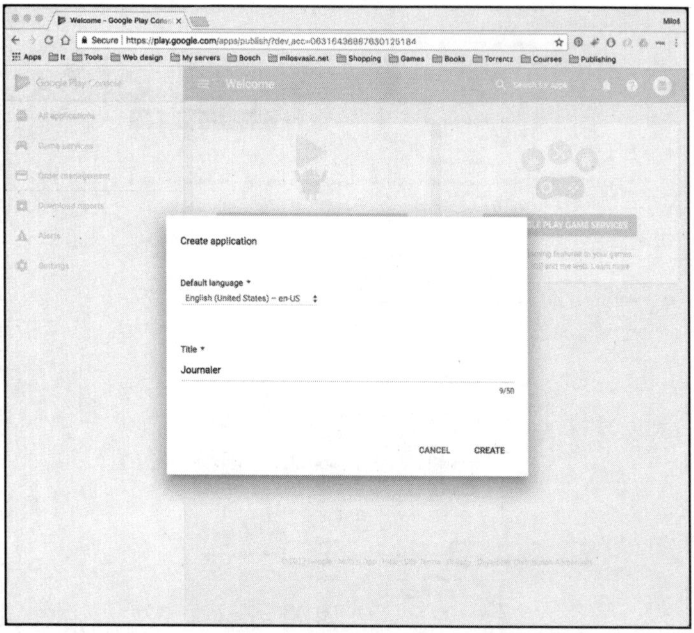

图 16.7

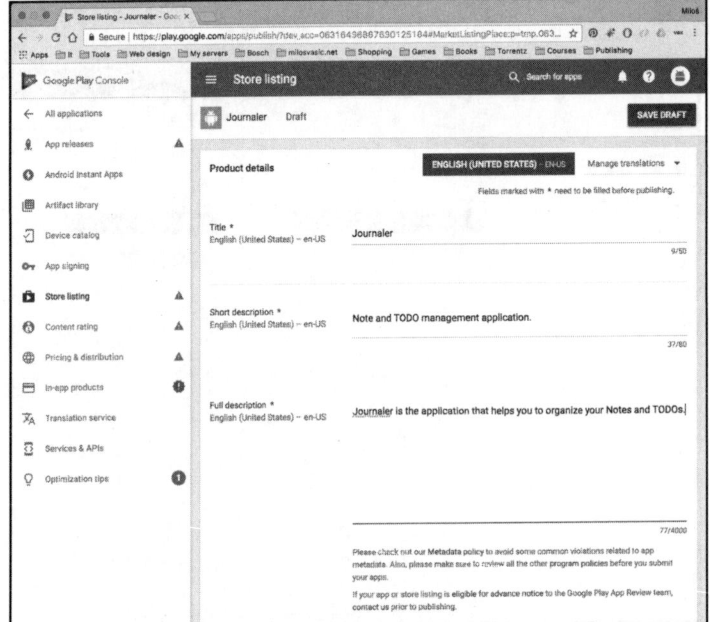
Store listing
Google Play Console
All applications
App releases
Android Instant Apps
Artifact library
Device catalog
App signing
Store listing
Content rating
Pricing & distribution
In-app products
Translation service
Services & APIs
Optimization tips
Journaler
Draft
SAVE DRAFT
Product details
ENGLISH (UNITED STATES) – EN-US
Manage translations
Fields marked with * need to be filled before publishing.
Title *
English (United States) – en-US
Journaler
9/50
Short description *
English (United States) – en-US
Note and TODO management application.
37/80
Full description *
English (United States) – en-US
Journaler is the application that helps you to organize your Notes and TODOs.
77/4000

图 16.8

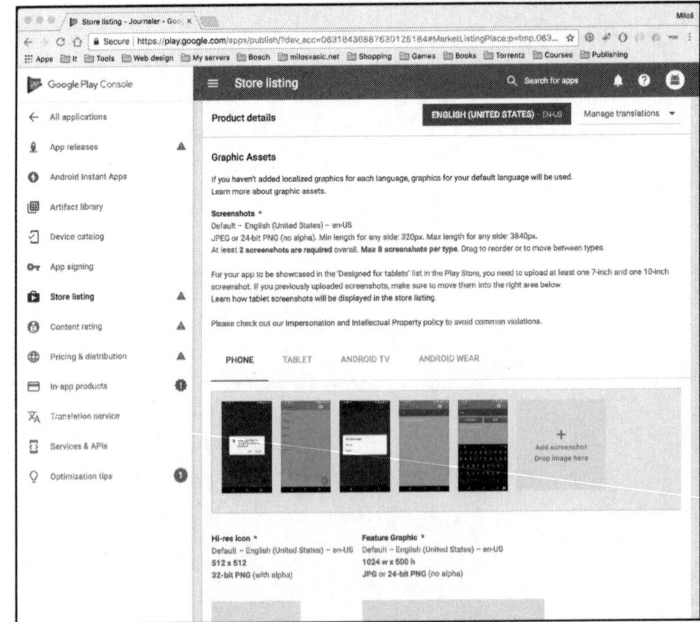
Store listing
Google Play Console
All applications
App releases
Android Instant Apps
Artifact library
Device catalog
App signing
Store listing
Content rating
Pricing & distribution
In-app products
Translation service
Services & APIs
Optimization tips
Product details
ENGLISH (UNITED STATES) – EN-US
Manage translations
Graphic Assets
Screenshots *
PHONE
TABLET
ANDROID TV
ANDROID WEAR
Add screenshot
Drop image here
Hi-res icon *
512 x 512
32-bit PNG (with alpha)
Feature Graphic *
1024 w x 500 h
JPG or 24-bit PNG (no alpha)

图 16.9

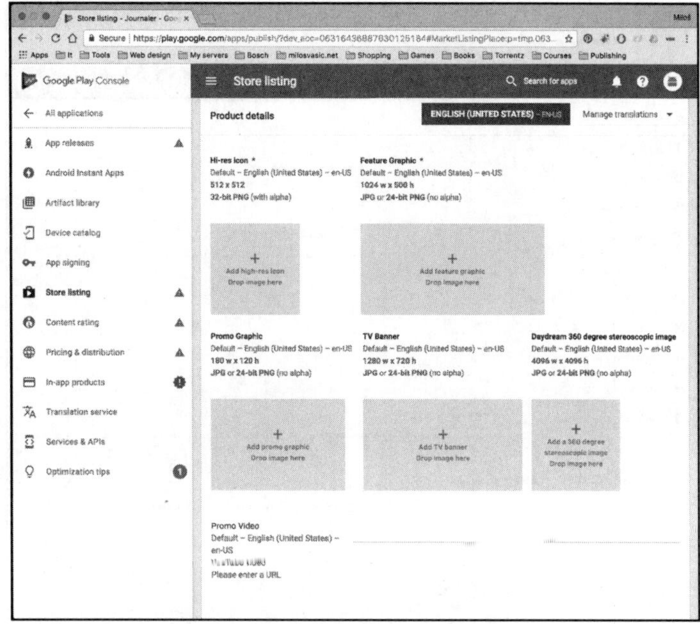
Store listing - Journaler
Google Play Console
Store listing
Search for apps
All applications
App releases
Android Instant Apps
Artifact library
Device catalog
App signing
Store listing
Content rating
Pricing & distribution
In-app products
Translation service
Services & APIs
Optimization tips
Product details
ENGLISH (UNITED STATES)
Manage translations
Hi-res icon *
Default – English (United States) – en-US
512 x 512
32-bit PNG (with alpha)
Feature Graphic *
Default – English (United States) – en-US
1024 w x 500 h
JPG or 24-bit PNG (no alpha)
Promo Graphic
Default – English (United States) – en-US
180 w x 120 h
JPG or 24-bit PNG (no alpha)
TV Banner
Default – English (United States) – en-US
1280 w x 720 h
JPG or 24-bit PNG (no alpha)
Daydream 360 degree stereoscopic image
Default – English (United States) – en-US
4096 w x 4096 h
JPG or 24-bit PNG (no alpha)
Promo Video
Default – English (United States) – en-US
Please enter a URL

图 16.10

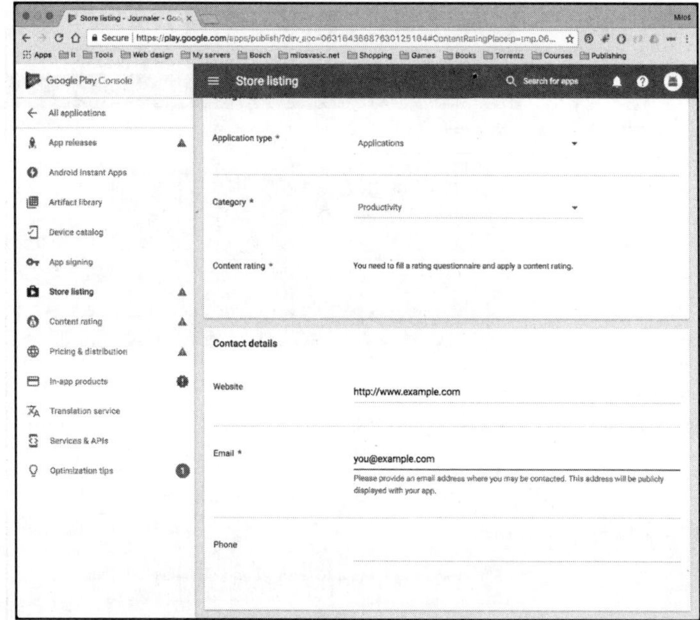
Store listing - Journaler
Google Play Console
Store listing
Search for apps
All applications
App releases
Android Instant Apps
Artifact library
Device catalog
App signing
Store listing
Content rating
Pricing & distribution
In-app products
Translation service
Services & APIs
Optimization tips
Application type *
Applications
Category *
Productivity
Content rating *
You need to fill a rating questionnaire and apply a content rating.
Contact details
Website
http://www.example.com
Email *
you@example.com
Please provide an email address where you may be contacted. This address will be publicly displayed with your app.
Phone

图 16.11

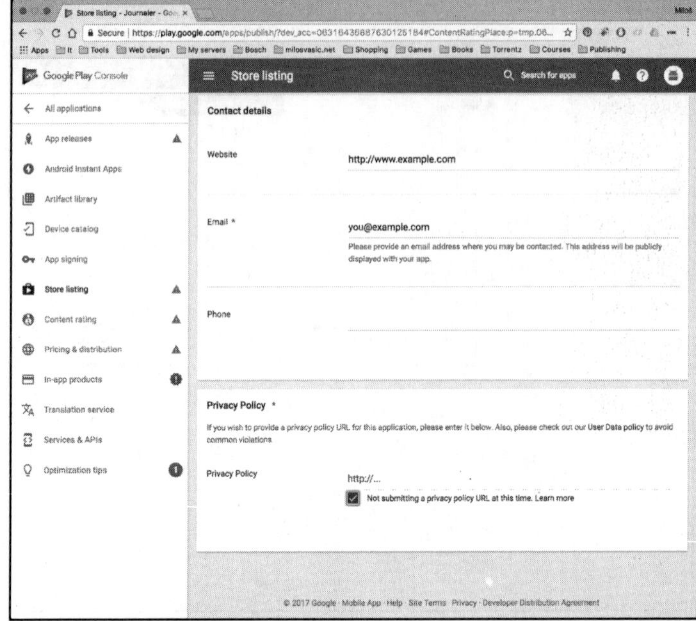
Store listing - Journaler - Goo
Google Play Console
Store listing
Search for apps
All applications
App releases
Android Instant Apps
Artifact library
Device catalog
App signing
Store listing
Content rating
Pricing & distribution
In-app products
Translation service
Services & APIs
Optimization tips
Contact details
Website
http://www.example.com
Email *
you@example.com
Please provide an email address where you may be contacted. This address will be publicly displayed with your app.
Phone
Privacy Policy *
If you wish to provide a privacy policy URL for this application, please enter it below. Also, please check out our User Data policy to avoid common violations.
Privacy Policy
http://...
Not submitting a privacy policy URL at this time. Learn more
© 2017 Google · Mobile App · Help · Site Terms · Privacy · Developer Distribution Agreement

图 16.12

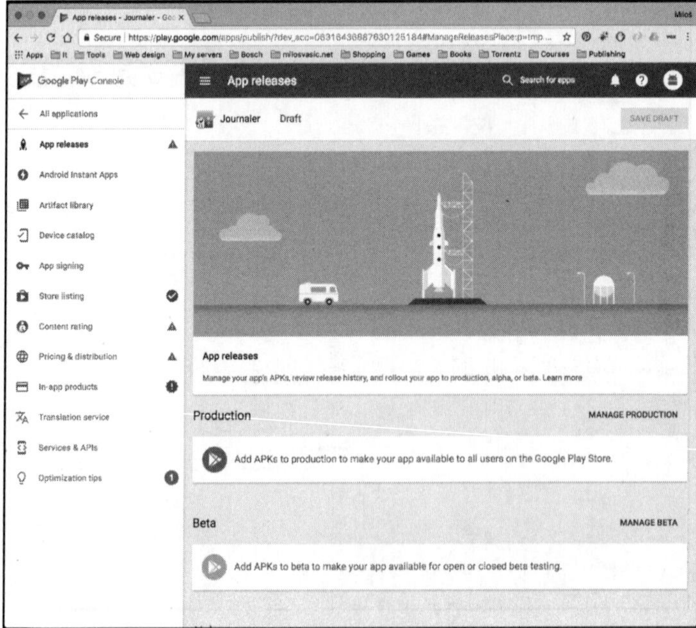
App releases - Journaler - Goo
Google Play Console
App releases
Search for apps
All applications
App releases
Android Instant Apps
Artifact library
Device catalog
App signing
Store listing
Content rating
Pricing & distribution
In-app products
Translation service
Services & APIs
Optimization tips
Journaler
Draft
SAVE DRAFT
App releases
Manage your app's APKs, review release history, and rollout your app to production, alpha, or beta. Learn more
Production
MANAGE PRODUCTION
Add APKs to production to make your app available to all users on the Google Play Store.
Beta
MANAGE BETA
Add APKs to beta to make your app available for open or closed beta testing.

图 16.13

此时将涵盖以下 3 个选项：

- ❑ MANAGE PRODUCTION。
- ❑ MANAGE BETA。
- ❑ MANAGE ALPHA。

取决于发布版本，可选择最适宜的选项。此处将选择 MANAGE PRODUCTION，然后单击 CREATE RELEASE 按钮，如图 16.14 所示。

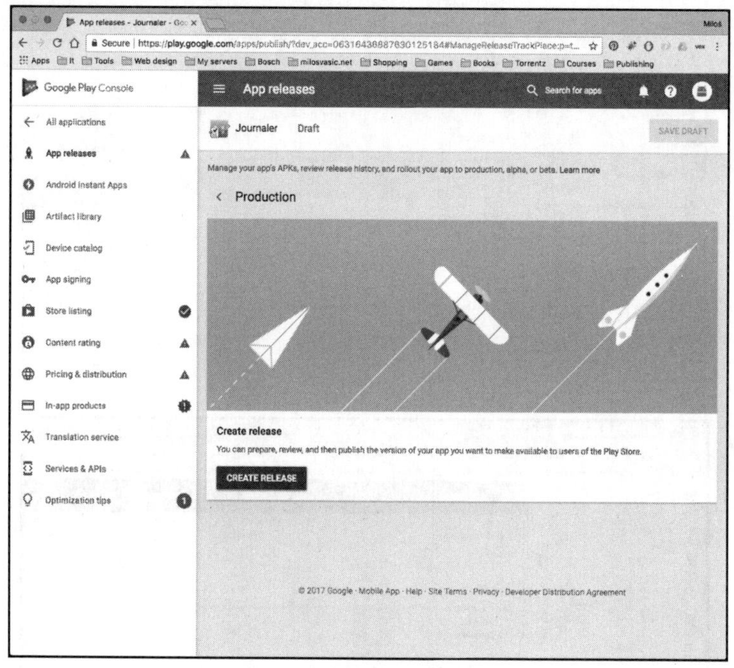

图 16.14

填充与发布版本相关的数据，如图 16.15 所示。

添加最近生成的 APK，并在页面下方处填充表单的其余部分。完成后单击 REVIEW 按钮查看应用程序发布版本，如图 16.16 所示。

在到达最终产品之前，单击左侧的 Content rating 链接，随后单击 CONTINUE 按钮，如图 16.17 所示。

填写 Email address，滚动至页面下方并选取分类，如图 16.18 所示。

选择 UTILITY, PRODUCTIVITY, COMMUNICATION, OR OTHER，则出现如图 16.19 所示的页面，在该页面中填写相关信息。

保存填写内容并单击 APPLY RATING 按钮，如图 16.20 所示。

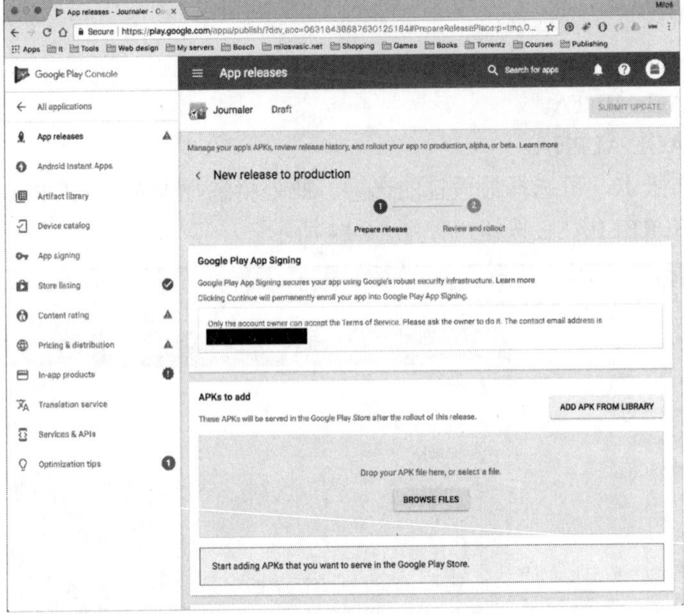

图 16.15

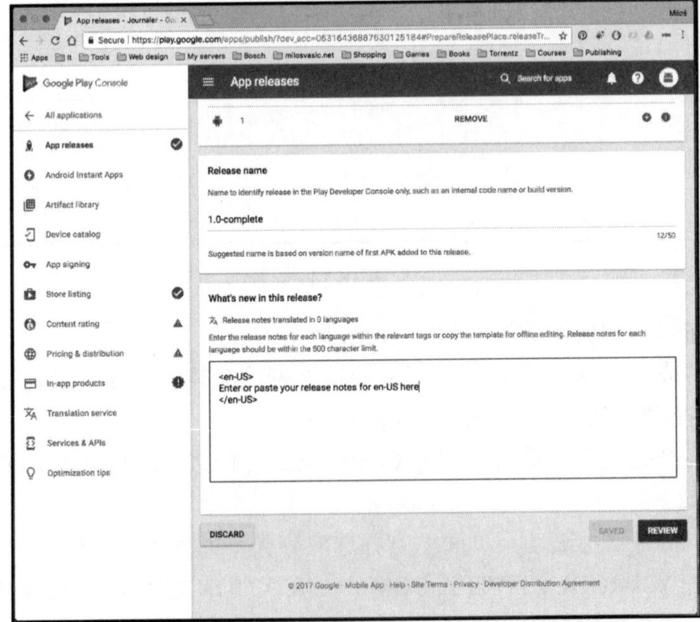

图 16.16

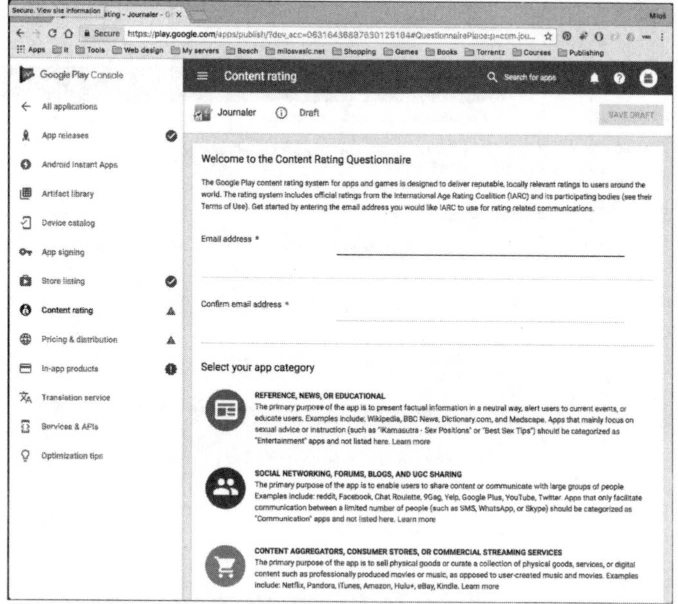
Google Play Console
All applications
App releases
Android Instant Apps
Artifact library
Device catalog
App signing
Store listing
Content rating
Pricing & distribution
In-app products
Translation service
Services & APIs
Optimization tips
Content rating
Journaler
Draft
SAVE DRAFT
Welcome to the Content Rating Questionnaire
Email address *
Confirm email address *
Select your app category
REFERENCE, NEWS, OR EDUCATIONAL
SOCIAL NETWORKING, FORUMS, BLOGS, AND UGC SHARING
CONTENT AGGREGATORS, CONSUMER STORES, OR COMMERCIAL STREAMING SERVICES

图 16.17

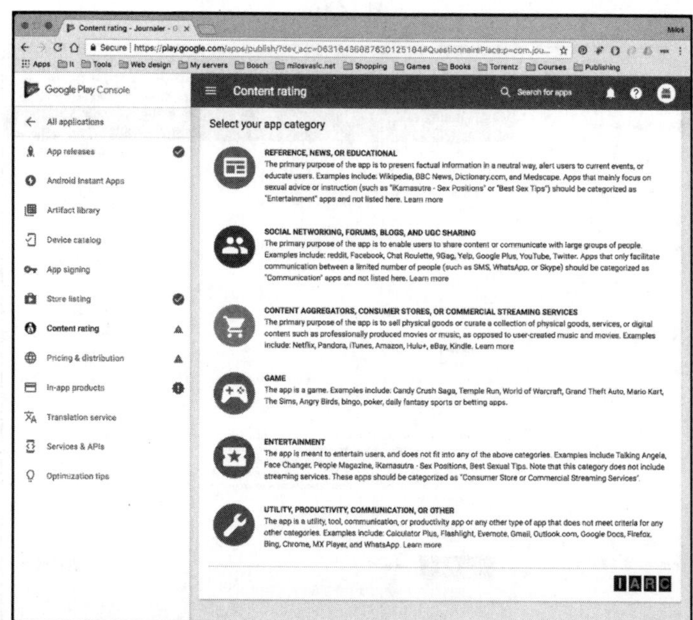
Google Play Console
All applications
App releases
Android Instant Apps
Artifact library
Device catalog
App signing
Store listing
Content rating
Pricing & distribution
In-app products
Translation service
Services & APIs
Optimization tips
Content rating
Select your app category
REFERENCE, NEWS, OR EDUCATIONAL
SOCIAL NETWORKING, FORUMS, BLOGS, AND UGC SHARING
CONTENT AGGREGATORS, CONSUMER STORES, OR COMMERCIAL STREAMING SERVICES
GAME
ENTERTAINMENT
UTILITY, PRODUCTIVITY, COMMUNICATION, OR OTHER
IARC

图 16.18

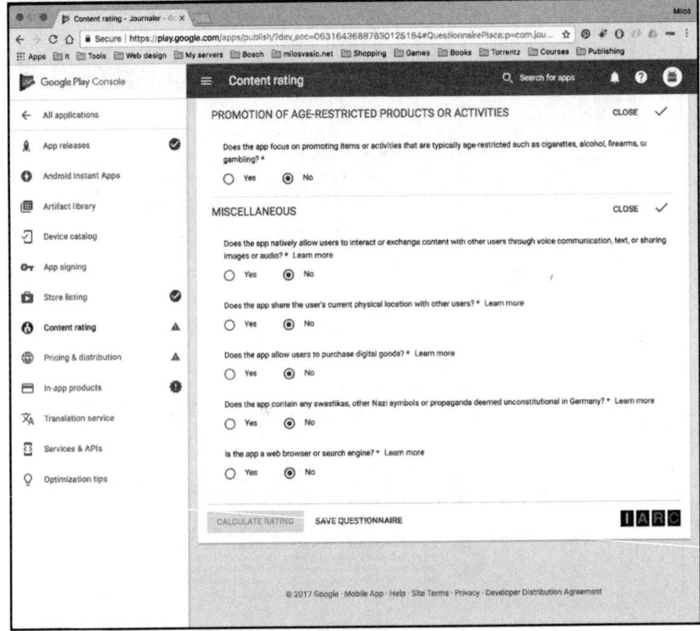
Content rating - Journaler
Secure | https://play.google.com/apps/publish/?dev_acc=06316436887830125184#QuestionnairePlace:p=com.jou...
Google Play Console
Content rating
All applications
App releases
Android Instant Apps
Artifact library
Device catalog
App signing
Store listing
Content rating
Pricing & distribution
In-app products
Translation service
Services & APIs
Optimization tips
PROMOTION OF AGE-RESTRICTED PRODUCTS OR ACTIVITIES
CLOSE
Does the app focus on promoting items or activities that are typically age-restricted such as cigarettes, alcohol, firearms, or gambling? *
Yes
No
MISCELLANEOUS
CLOSE
Does the app natively allow users to interact or exchange content with other users through voice communication, text, or sharing images or audio? * Learn more
Does the app share the user's current physical location with other users? * Learn more
Does the app allow users to purchase digital goods? * Learn more
Does the app contain any swastikas, other Nazi symbols or propaganda deemed unconstitutional in Germany? * Learn more
Is the app a web browser or search engine? * Learn more
CALCULATE RATING
SAVE QUESTIONNAIRE
IARC
© 2017 Google · Mobile App · Help · Site Terms · Privacy · Developer Distribution Agreement

图 16.19

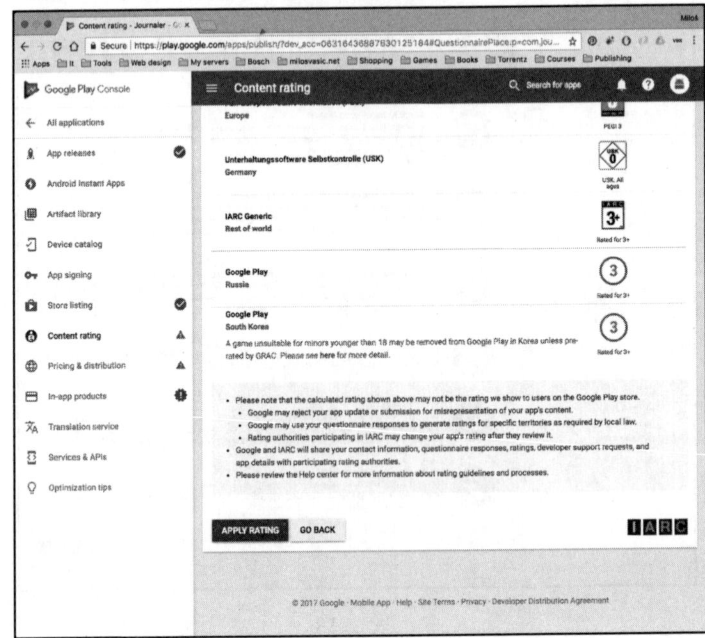
Content rating - Journaler
Google Play Console
Content rating
Europe
PEGI 3
Unterhaltungssoftware Selbstkontrolle (USK)
Germany
USK All ages
IARC Generic
Rest of world
Rated for 3+
Google Play
Russia
Google Play
South Korea
A game unsuitable for minors younger than 18 may be removed from Google Play in Korea unless pre-rated by GRAC. Please see here for more detail.
Please note that the calculated rating shown above may not be the rating we show to users on the Google Play store.
Google may reject your app update or submission for misrepresentation of your app's content.
Google may use your questionnaire responses to generate ratings for specific territories as required by local law.
Rating authorities participating in IARC may change your app's rating after they review it.
Google and IARC will share your contact information, questionnaire responses, ratings, developer support requests, and app details with participating rating authorities.
Please review the Help center for more information about rating guidelines and processes.
APPLY RATING
GO BACK
IARC
© 2017 Google · Mobile App · Help · Site Terms · Privacy · Developer Distribution Agreement

图 16.20

切换至 Pricing & distribution 部分，如图 16.21 所示。

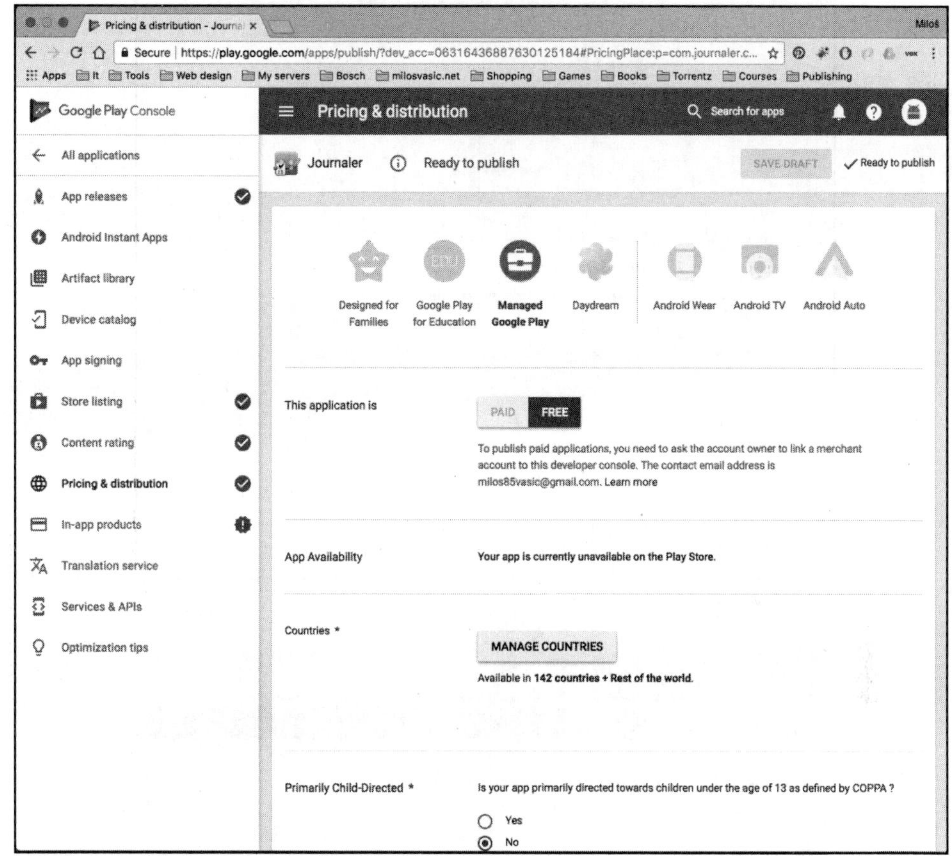

图 16.21

在图 16.21 中，填写表单并设置所询问的相关数据。完成后单击页面上方的 SAVE DRAFT 按钮，此时会出现 Ready to publish 链接，单击该链接，则出现如图 16.22 所示的页面。

单击 MANAGE RELEASES 按钮后，直至到达 App releases 部分中的最后一个页面，即可看到 START ROLLOUT TO PRODUCTION 按钮处于启用状态，如图 16.23 所示。单击该按钮，并在后续询问过程中单击 CONFIRM 按钮。

继续执行当前操作，如图 16.24 所示。

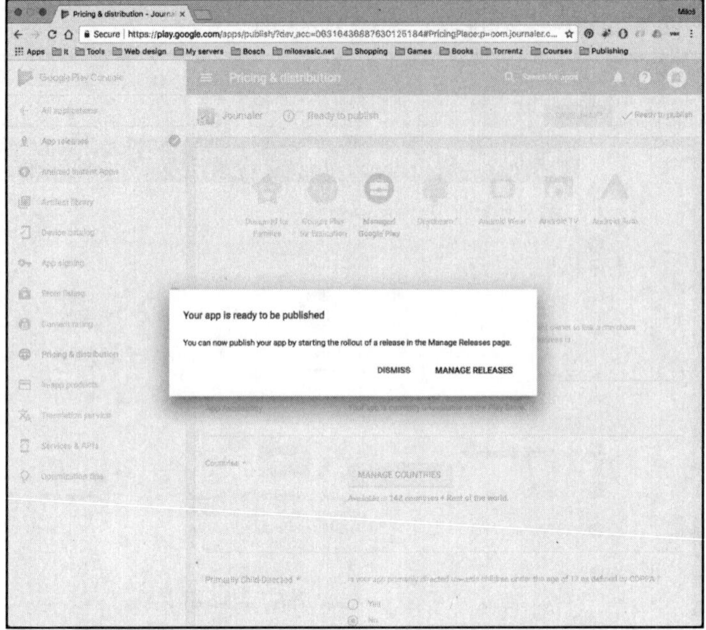
Your app is ready to be published
You can now publish your app by starting the rollout of a release in the Manage Releases page.
DISMISS
MANAGE RELEASES

图 16.22

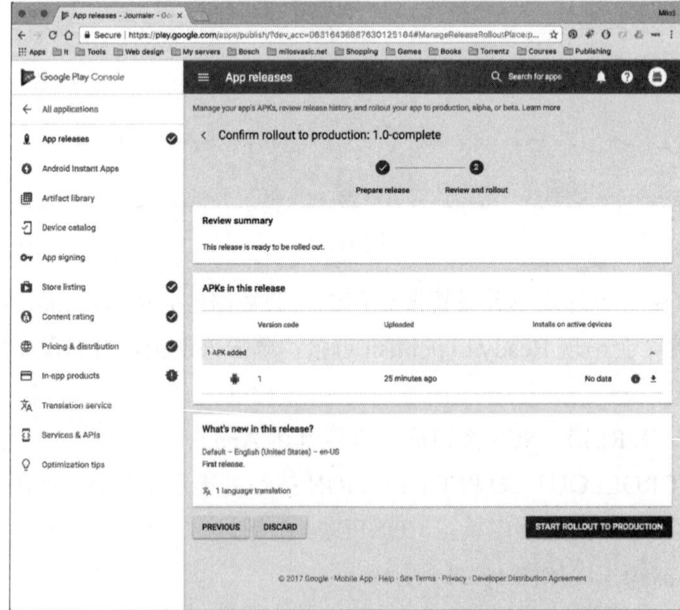
Google Play Console
App releases
All applications
App releases
Android Instant Apps
Artifact library
Device catalog
App signing
Store listing
Content rating
Pricing & distribution
In-app products
Translation service
Services & APIs
Optimization tips
Confirm rollout to production: 1.0-complete
Prepare release
Review and rollout
Review summary
This release is ready to be rolled out.
APKs in this release
Version code
Uploaded
Installs on active devices
1 APK added
1
25 minutes ago
No data
What's new in this release?
Default – English (United States) – en-US
First release.
1 language translation
PREVIOUS
DISCARD
START ROLLOUT TO PRODUCTION

图 16.23

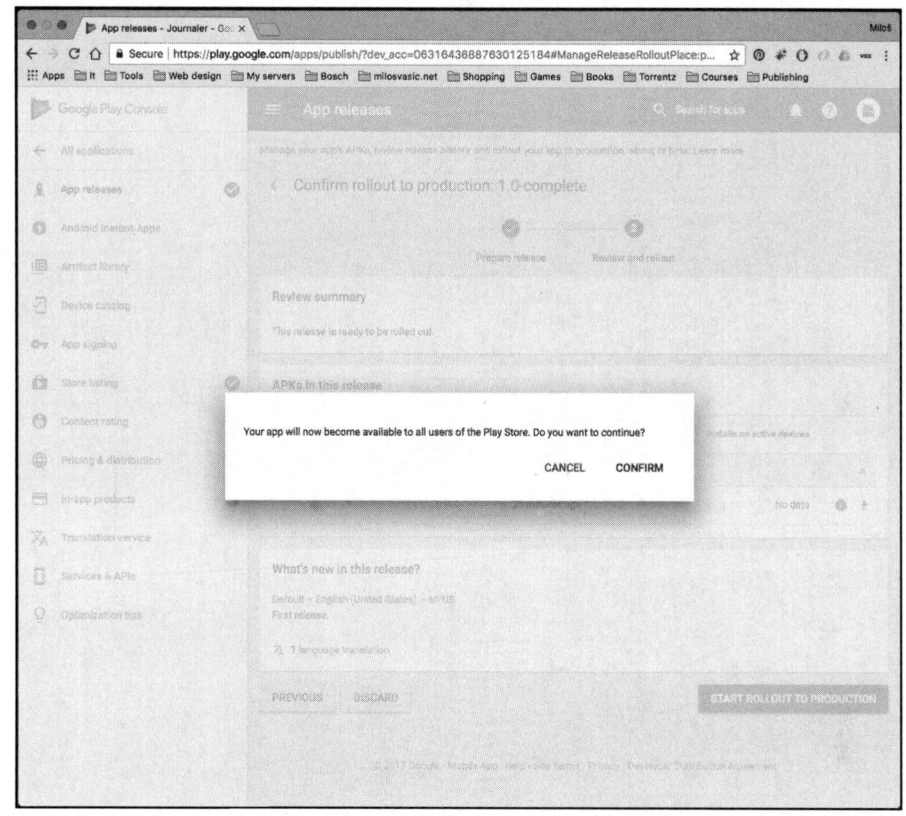

图 16.24

至此，我们已经成功地向 Google Play store 中发布了应用程序。

16.5　本 章 小 结

本章逐步介绍了部署的处理过程，其中涉及了大量的工作。

接下来，读者应仔细考查所要构建的应用程序，并从头开始加以设计。在开发过程中，读者还会面临一些之前未涉及的问题。Android 中涵盖了大量的内容，了解整个框架可能会消耗大量的时间，而许多开发人员并不了解其中的每个部分。读者应尽可能地编写代码，进而深入理解通过本书所学到的内容。祝您好运！

[illegible] Google Play [illegible]

本章小结

[illegible]